THE WONDERS OF THE HUMAN BODY

THE WONDERS OF THE HUMAN BODY

PHYSICAL REGENERATION ACCORDING THE LAWS OF CHEMISTRY AND PHYSIOLOGY

GEORGE W. CAREY

GYAN

Published by

Gyan Publishing House
5, Ansari Road
Daryaganj, New Delhi-110002
Phone: 011-47034999, 9811692060
E-mail: books@gyanbooks.com

Distribution Network
gyanbooks.com
India, USA, Canada, UK, Australia

ISBN: 978-93-7169-435-3 (HB)

First Published, 1918

2nd Impression 2025

Printed at: Gyan Press, Delhi.

THE WONDERS OF THE HUMAN BODY
Author: GEORGE W. CAREY

ADVENT SEPT. 7, 1845

Faithfully Yours.
George W. Carey.

The Wonders of the Human Body

Physical Regeneration
According to the Laws of
Chemistry and Physiology

by

DR. GEORGE W. CAREY
Author of Text Books
on
The Chemistry of Life

Published by
THE CHEMISTRY OF LIFE CO.
Los Angeles, Cal.
1918

DEDICATION

To the Spirit of the Declaration of Independence;

To the three great pilots, George Washington, Abraham Lincoln and Woodrow Wilson;

To the noble, brave American soldiers who are upholding World-Democracy on the battlefields of Europe; and

To the self-sacrificing women—the mothers of America who stand before almighty God and give all they have to transmit to posterity a world free and redeemed from autocracy—this book is reverently dedicated.

INTRODUCTION

THE days of speculative philosophy have ended. One by one, those who have been blind are commencing to see; ears that were deaf are bginning to hear, and more and more people are recognizing the importance and the significance of the ancient command, "Man, know thyself."

Hungry to learn, the earnest investigator, with eager intellect and open, unbiased mind, knocks at the door of the sanctuary and asks: "*Where* is the Kingdom of Heaven?"

"What is salvation?"

"*How* can I be saved?"

The answers hitherto given are indefinite and vague.

The laboratory door opens and a Professor greets the anxious enquirer. "I cannot explain to you the mysteries of life here," he said, "or that of the life hereafter, if, indeed, we live hereafter. We hold certain theories and can tell you somewhat of the composition of your body; but we know not where the Kingdom of Heaven is, nor the actual process by which salvation may be attained."

And thus neither the church nor the laboratory have solved the age-long problem. They have blazed no trail to the Kingdom of Heaven; neither have they analyzed soul-stuff, or located the Holy of Holies.

Is there, then, *any* solution to these problems?

Where can the Kingdom be found?

And if found, anywhere, why has not the church and the laboratory discovered it?

The church could have found it. The laboratory could have found it.

"Seek and ye shall find"; "Knock and it shall be opened unto you".

The church has not sought God; neither has the laboratory.

The churches and the laboratories have sought *mammon.*

They *tried* to make themselves believe that they were searching for God, and sometimes they succeeded in making others believe it, also.

If they had truly and humbly sought, they would have found that which they sought.

Does the Almighty make a promise and not keep it?

The Bible, the Kaballah, the Vedas, and ancient cuneiform tablets, torn from their hiding places in the heart of Mother Nature, all hold the key to the wisdom of the ages; and some there are who have turned the key and the door to understanding has opened.

The author of this work has found the key and he has turned it. Erect and unafraid—for he has nothing to lose or gain—he reverently offers you that which you are seeking.

He does not claim to be the only one who has discovered the key, but, working along the lines hinted at by some modern and many ancient seekers, he hereby presents the results of his many years of scientific research.

He shows that the correct translations of the Greek and Hebrew scriptures plainly state that all the allegories and fables of these works refer to actual physiological facts; and the statements—"Seek ye first the Kingdom of Heaven and all things shall be added unto you"; "The Kingdom of Heaven is within you", and "Be ye therefore perfect as your Father in Heaven is perfect" can be physiologically and chemically explained.

INEZ E. PERRY.

FOREWORD

June 1st, 1917, Dr. Carey published the Tree of Life, a book on the marvels of the human body and physical regeneration as taught in Greek and Hebrew texts, from which the English Bible was translated.

The book aroused such intense interest that the edition was virtually sold out in less than a year.

The Wonders of the Human Body may be termed "an enlarged edition of the Tree of Life", containing more than double the amount of information.

Many more Bible witnesses have been called to the stand, than were heard in the Tree of Life, to testify to the truths of Scripture and to indubitably prove that the Greek and Hebrew writings were based in the truths of chemistry and physiology.

This book was not compiled for the purpose of antagonizing so-called Christians. It was written to prove the truths of the fables, parables and splendid allegories, such as the book of Job and Revelations, with a firm faith that truth will triumph and at last lead the world to Peace.

THE AUTHOR.

Los Angeles, California, September 7, 1918.

Table of Contents

PART ONE

PART TWO

PART ONE

THE WONDERS OF THE HUMAN BODY

> "For thou didst cover me in my mother's womb.
> I will give thanks unto thee; for I am fearfully and wonderfully made;
> Wonderful are thy works."
>
> —*139th Ps., 13 and 14th v.*

THE human race has been asleep, and has dreamed that property and money are the true wealth of a nation, sacrificing men, women and children to the chimerical idea that danced in visionary splendor through their brains. The result of this is to be seen in the uneasiness that prevails everywhere. But humanity is waking up, slowly but surely, and beginning to realize that it, itself, is the most precious thing on earth.

The old-established statement that the individuals that make up the race are imperfect is no more true than that a pile of lumber is imperfect, that is to be afterward reformed, or built into a house. As it is the carpenter's business to take the lumber, which is perfect as material, and build the house, so it is the legitimate work of spiritual man to take the perfect material everywhere present and build, by the perfect law of chemistry and mathematics, the perfected, harmonious human being, and, with this material, employ the same law to build up society collectively.

It is a well-known physiological fact that the blood is the basic material of which the human body is continually builded. As is the blood, so is the body; as is the body, so is the brain; as is the brain, so is the quality of thought. As a man is builded, so thinks he.

According to the views of students of modern alchemy, the Bible—both the Old and the New Testaments—are symbolical writings, based primarily upon this very process of body building. The word alchemy really means fleshology. It is derived from chem, an ancient Egyptian word, meaning flesh. The word Egypt also means flesh, or anatomy.

Alchemy, however, in its broader scope, means the science of solar rays. Gold may be traced to the sun's rays. The word gold means solar essence. The transmutation of gold does not mean the process of making gold, but does mean the process of changing gold, solar rays, into all manner of materialized forms, vegetable, mineral, etc. The ancient alchemist studied the process of Nature in her operations from the volatile to the fixed, the fluid to the solid, the essence to the substance, or the abstract to the concrete, all of which may be summed up in the changing of spirit into matter. In reality, the alchemist did not try to do anything. He simply tried to search out nature's processes in order that he might comprehend her marvelous operations.

To be sure, language was used that to us seems symbolical and often contradictory, but it was not so intended, nor so at all in reality. We speak in symbols. If a man is in delirium, caused by alcohol in his brain-cells, we say he has "snakes in his boots." Of course, no one supposes that the words are to be taken literally. Yet, if our civilization should be wiped out, and our literature translated after four or five thousand years, those who read our history might be puzzled to know what was meant by "snakes in his boots."

Again, it has been believed by most people that the words, "transmutation of base metals into gold," used by alchemists, referred to making gold. But a careful study of the Hebrew Cosmogony, and the Kabala, will reveal the fact that the alchemist always referred to solar rays when he used the word gold. By "base metals," they simply meant matter, or basic material. The dissolving or disintegration of matter, the combustion of wood or coal, seemed as wonderful to these philosophers as the growth of wood or the formation of coal or stone. So the transmutation of base metals into gold simply meant the process of changing the fixed into the volatile, or the dematerialization of matter, either by heat or chemical process.

It is believed by modern students of alchemy that the books of the Old and New Testaments are a collection of alchemical and astrological writings, dealing entirely with the wonderful operation of aerial elements (spirit) in the human body, so fearfully and wonderfully made. The same authority is given for the statement, "Know ye not that your bodies are the temple of the living God" and "Come unto Me all ye that labor and are heavy laden and I will give you rest." According to the method of reading the numerica' value of letters by the Kabala, M and E figure B, when united. Our B is from the Hebrew Beth, meaning a house or temple—the temple of the spiritual ego—the body. Thus by coming into the realization that the body is really the Father's House, temple of God, the spirit secures peace and contentment or rest.

The human body is composed of perfect principles, gases, minerals, molecules, or atoms; but these builders of flesh and bone are not always properly adjusted. The planks or bricks used in building houses may be endlessly diversified in arrangement, and yet be perfect material.

Solomon's temple is an allegory of man's temple—the human organism. This house is built (always being built) "without sound of saw or hammer."

The real Ego manifests in a house, beth, church, or temple—i. e., Soul-of-Man's Temple, for the Ego or I AM.

The solar (soular) plexus is the great central sun or dynamo on which the Subconscious Mind (another name for God) operates and causes the concept of individual consciousness. Specifically stated thus:

1. The upper brain (cerebrum); "The Most High" or Universal Father, which furnishes substance for all functions that constitute the body.

2. The Spiritual Ego ("I AM") resident in the cerebellum.

3. The Son of God, the redeeming seed or Jesus, born monthly in the solar plexus.

4. Soul, the fluids of the body.

5. Flesh, bone, etc., the fluids materialized. (In a broader sense body also is termed soul, "Every soul per-

ished.") It is not thinkable that every Spirit, or Ego or "I AM" was drowned.

No wonder that the seers and alchemists of old declared that "Your bodies are the temple of the living God" and "The kingdom of Heaven is within you." But man, blinded by selfishness, searches here and there, scours the heavens with his telescope, digs deep into earth, and dives into ocean's depths, in a vain search for the elixir of life that may be found between the soles of his feet and the crown of his head. Really our human body is a miracle of mechanism. No work of man can compare with it in accuracy of its process and the simplicity of its laws.

At maturity, the human skeleton contains about 165 bones, so delicately and perfectly adjusted that science has despaired of ever imitating it. The muscles are about 500 in number; length of alimentary canal, 32 feet; amount of blood in average adult, 30 pounds, or one-fifth the weight of the body; the heart is six inches in length and four inches in diameter, and beats seventy times per minute, 4200 times per hour, 100,800 per day, 36,720,000 per year. At each beat, two and one-half ounces of blood are thrown out of it, 175 ounces per minute, 656 pounds per hour, or about eight tons per day.

All the blood in the body passes through the heart every three minutes; and during seventy years it lifts 270,000,000 tons of blood.

The lungs contain about one gallon of air at their usual degree of inflation. We breathe, on an average, 1200 breaths per hour; inhale 600 gallons of air, or 24,000 gallons daily.

The aggregate surface of the air-cells of the lungs exceeds 20,000 square inches, an area nearly equal to that of a room twelve feet square. The average weight of the brain of an adult is three pounds, eight ounces; the average female brain, two pounds, four ounces. The convolutions of a woman's brain cells and tissues are finer and more delicate in fibre and mechanism, which evidently accounts for the intuition of women. It would appear that the difference in the convolutions and fineness of tissue in brain matter is responsible for the degrees of consciousness called reason and intuition.

The nerves are all connected with the brain directly, or by the spinal marrow, but nerves receive their sustenance from the blood, and their motive power from the solar plexus dynamo. The nerves, together with the branches and minute ramifications, probably exceed ten millions in number, forming a bodyguard outnumbering the mightiest army ever marshalled.

The skin is composed of three layers, and varies from one-eighth to one-quarter of an inch in thickness. The average area of skin is estimated to be about 2000 square inches. The atmospheric pressure, being fourteen pounds to the square inch, a person of medium size is subject to a pressure of 40,000 pounds. Each square inch of skin contains 3500 sweat tubes, or perspiratory pores (each of which may be likened to a little drain tile) one-fourth of an inch in length, making an aggregate length of the entire surface of the body of 201,166 feet, or a tile for draining the body nearly forty miles in length.

Our body takes in an average of five and a half pounds of food and drink each day, which amounts to one ton of solid and liquid nourishment annually, so that in seventy years a man eats and drinks 1000 times his own weight.

There is not known in all the realms of architecture or mechanics one little device which is not found in the human organism. The pulley, the lever, the inclined plane, the hinge, the "universal joint," tubes and trap-doors; the scissors, grind-stone, whip, arch, girders, filters, valves, bellows, pump, camera, and Aeolian harp; and irrigation plant, telegraph and telephone systems—all these and a hundred other devices which man thinks he has invented, but which have only been telegraphed to the brain from the Solar Plexus (cosmic centre) and crudely copied or manifested on the objective canvas.

No arch ever made by man is as perfect as the arch formed by the upper ends of the two legs and the pelvis to support the weight of the trunk. No palace or cathedral ever built has been provided with such a perfect system of arches and girders.

No waterway on earth is so complete, so commodious, or so populous as that wonderful river of life, the "Stream of Blood." The violin, the trumpet, the harp, the grand organ, and all the other musical instruments, are mere counterfeits of the human voice.

Man has tried in vain to duplicate the hinges of the knee, elbow, fingers and toes, although they are a part of his own body.

Another marvel of the human body is the self-regulation process by which nature keeps the temperature in health at 98 degrees. Whether in India, with the temperature at 130 degrees, or in the arctic regions, where the records show 120 degrees below the freezing point, the temperature of the body remains the same, practically steady at 98 degrees, despite the extreme to which it is subjected.

It was said that "All roads lead to Rome." Modern science has discovered that all roads of real knowledge lead to the human body. The human body is an epitome of the universe; and when man turns within the mighty searchings of reason and investigation that he has so long used without—the "New Heaven and Earth" will appear.

While it is true that flesh is made by a precipitation of blood, it is not true that blood is made from food. The inorganic or cell-salts contained in food are set free by the process of combustion or digestion, and carried into the circulation through the delicate absorbent tubes of the mucous membrane of stomach and intestines. Air, or Spirit, breathed into the lungs, enters the arteries (air carriers) and chemically unites with the mineral base, and by a wonderful transformation creates flesh, bone, hair, nails, and all the fluids of the body.

On the rock (Peter or Petra, meaning stone) of the mineral salts is the human structure built, and the grave, stomach, or hell shall not prevail against it. The minerals in the body do not disintegrate or rot in the grave.

The fats, albumen, fibrine, etc., that compose the organic part of food, are burned up in the process of digestion and transposed into energy or force to run the human battery. Blood is made from air; thus all nations that dwell on earth are of one blood, for all breathe one air. The best food is the food that burns up quickest and easiest; that is, with the least friction in the human furnace.

The sexual functions of man and woman; the holy operation of creative energy manifested in male and female; the formation of life germs in ovum and sex fluids; the Divine Procedure of the "word made flesh" and the mysteries of conception and birth are the despair of science.

"Know ye not that your bodies are the temple of the living God?" for "God breathed into man the breath of life."

In the words of Epictitus, "Unhappy man, thou bearest a god with thee, and knoweth it not."

Walt Whitman sings:

"I loaf and invite my soul; I lean and loaf at my ease, observing a spear of summer grass. Clean and sweet is my soul, and clean and sweet is all that is not my soul."

"Welcome every organ and attribute of me, and of any man hearty and clean, not an inch, not a particle of an inch, is vile, and none shall be less familiar than the rest."

"Divine am I, inside and out, and I make holy whatever I touch or am touched from."

"I say no man has ever yet been half devout enough; none has ever yet adored or worshipped half enough; none has begun to think how divine he himself is, and how certain the future is."

The vagus nerve, so named because of its wandering (vagrant) branches, is the greatest marvel of the human organism. Grief depresses the circulation through the vagus, a condition of malnutrition follows, and tuberculosis, often of the hasty type, follows.

The roots of the vagus nerve are in the medulla oblongata, at the base of the small brain or cerebellum, and explains why death follows the severing of the medulla. It controls the heart action, and if a drug such as aconite be administered, even in small doses, its effect upon this nerve is shown in slowing the action of the heart and decreasing the blood pressure. In larger doses it paralyzes the ends of the vagus in the heart, so that the pulse becomes suddenly very rapid and at the same time irregular. Branches of the vagus nerve reach the heart, lungs, stomach, liver and kidneys.

Worry brings on kidney disease, but it is the vagus nerve, and especially that branch running to the kidneys which undue excitement or worry, or strain, brings about the paralysis of the kidneys in the performance of their functions.

When we say that a man's heart sinks within him from fear or apprehension, it is shown by the effect of this nerve upon the heart action. If his heart beats high with hopes, or he sighs for relief, it is the vagus nerve that has conducted the mental state to the heart and accelerated its action or caused that spasmodic action of the lungs which we call a sigh.

The nerves of the human body constitute the "Tree of Life," with its leaves of healing. The flowing waters of the "Rivers of Life" are the veins and arteries through which sweep the red, magnetic currents of Love—of Spirit made visible.

Behold the divine telegraph system, the million nerve wires running throughout the wondrous temple, the temple not made with hands, the temple made "without sound of saw or hammer." View the Central Sun of the human system—the Solar Plexus—vibrating life abundantly.

Around this dynamo of God, you may see the Beasts that worship before the Throne day and night saying, "Holy holy, art Thou, Lord God Almighty." The Beasts are the twelve plexuses of nerve centers, telegraph stations, like unto the twelve zodiacal signs that join hands in a fraternal circle across the gulf of space.

Aviation, liquified air, deep breathing for physical development and the healing of divers diseases rule the day. In every brain there are dormant cells, waiting for the "coming" of the bridegroom, the vibration of the air age (the Christ) that will resurrect them.

Everywhere we have evidence of the awakening of dormant brain cells. Much, if not all, of spiritual phenomena, multiple personality, mental telepathy, and kindred manifestations are explainable upon the hypothesis of the possibility of awakening and bringing into use of dormant brain cells.

Scientists have discovered that there are dormant, or undeveloped brain cells in countless number, especially in the cerebrum, or upper brain, the seat of the moral faculties; or, more definitely speaking, the key, which, when touched with the vital force set free through the process of physical regeneration by saving the seed and by the baptism in ointment (Christ) in the spinal cord, lifts the crucified substance wasted by the prodigal son in "riotous living" up to the "most high" brain.

This procedure causes the dormant cells, little buds, to *actually bloom.* The simile is perfect. The cells, while dormant, are like a flower yet in the bud. When the Substance that is needed for their development reaches them, through physical regeneration as outlined in the "Plan of Salvation," fully explained in part 2nd., the cells *bloom* and then vibrate at a rate that causes the Consciousness of the "New Birth."

"He that is born of God doth not sin, *for his seed remaineth in him." John.*

The eye is hardly less wonderful, being a perfect photographer's camera. The retina is the dry plate on which are focused all objects by means of the crystalline lens. The cavity behind this lens is the shutter. The eyelid is the drop shuttle. The draping of the optical dark room is the only black membrane in the entire body. The miniature camera is self-focusing, self-loading and self-developing, and takes millions of pictures every day in colors and enlarged to life size.

Charts have been prepared—marvelous charts—which go to show that the eye has 729 distinct expressions conveying as many distinct shades of meaning.

The power of color perception is overwhelming. To perceive red the retina of the eye must receive 395,000 vibrations in a second; for violet it must respond to 790,000,000. In our waking moments our eyes are bombarded every minute by at least 600,000,000 vibrations.

The ear contains a perfect miniature piano of about 3000 double fibers or strings stretched or relaxed in unison with exterior sounds. The longest cord of this marvelous instrument is one-fifteenth of an inch, while the shortest is about one-five-hundredth of an inch. The 3000 strings are distributed through a register of seven octaves, each octave corresponding to about 500 fibers and every half tone subdivided again into 320 others. The deepest tone we can hear has thirty-two vibrations a second, the highest has 70,000.

The ear is a colossal mystery, and the phenomenon of sound is a secret only recorded in the Holy of Holies of the Infinite Mind. And what is mind? We know absolutely nothing about it. Some believe that mind is the product of the chemical operation of matter, viz.: the atoms or materials that compose the human body. These persons contend that all electrons are particles of pure Intelligence and KNOW what to do. Others hold to the theory that universal Mind (whatever it may be) forms a body from some material, they know not what, and then plays upon it or operates through it.

Visions of beauty and splendor,
   Forms of a long-lost race,
Sounds of voices and faces
   From the fourth dimensions of space;
And on through the universe boundless,
   Our thoughts go, lightning-shod;
Some call it Imagination,
   And others call it God!

Now comes speech, the Word that was in the beginning. God certainly bankrupted His infinite series of miracles when He gave the power of speech to man.

We wonder and adore in the presence of that pulsing orb, the heart. Tons of the water of life made red by the Chemistry of Love sweep through this central throne every day, and flow on to enrich the Edenic Garden until its waste places shall bloom and blossom as the rose.

Take my hand and go with me to the home of the Spiritual Ego—the wondrous brain. Can you count the whirling, electric, vibrating cells? No, not until you can count the sand grains on the ocean's shore. These rainbow-hued cells are the keys that the fingers of the soul strike to play its part in the Symphony of the Spheres.

At last we have seen the "Travail of the Soul and are satisfied." No more temples of the Magi now, but instead the glorious human Beth. At last we have found the true church of God, the human body. In this body, or church, spirit operates like some wizard chemist or electrician. No more searching through India's jungles or scaling the Himalayan heights in search for a master — a mahatma — or ancient priest, dwelling in some mysterious cave where occult rites and ceremonies are supposed to reveal the wisdom of the past. But, instead, you have found the Kingdom of the Real within the Temple that needs no outer Sun by day nor Moon or Stars by night to lighten it. And then the enraptured Soul becomes conscious that the stone has been rolled away from the door of material concept where it has slept, and it now hears the voice of the Father within saying, "Let there be light!" and feels the freedom that comes with knowing that Being is one.

And now man also realizes the meaning of the "Day of Judgment." He realizes that Judgment means understanding, hence the ability to judge. He then judges correctly, for he sees the Wisdom of Infinite Life in all men, in all things, all events, and all environments. Thus does the new birth take place, and the Kingdom of Harmony reigns now.

Man must realize, however, that he is the creator or builder of his own body, and that he is responsible for every moment of its building, and every hour of its care. He alone can select and put together the materials provided by the universe for its construction. Man has been able to scale the heavens, to measure the distances between the stars and planetary bodies, and to analyze the component parts of suns and worlds, yet he cannot eat without making himself ill; he can fore-tell eclipses and tides for years in advance, but cannot look far enough ahead in his own affairs to say when he may be brought down with la grippe, or to calculate accurately the end of any bodily ill that may afflict him. When he finds out what he really is, and how much he has always had to do in the making of himself what he is, he will be ready to grasp some idea of the wonderful possibilities of every human body, and will know how completely and entirely is every man his own savior. Just so long as he denies his own powers, and looks outside of himself for salvation from present or future ills, he is indeed a lost creature. If the race is to be redeemed, it must come as the result of thought followed by action. If the race is to think differently than at present, it must have new bodies with new brains.

Modern physiologists know that our bodies are completely made over every year, by the throwing off of worn-out cells and the formation of new ones, that is going on every minute. Nature will take care of the making-over process, but we are responsible for the plan of reconstruction. Man must learn to run the machinery of his body with the same mathematical accuracy as he now displays in control of an engine or automobile, before he can lay claim to his divine heritage and proclaim himself master of his own.

The law of life is not a separate agent working independently of mankind and separate from individual life. Man himself is a phase of the great law in operation. When he once fully awakens to the universal co-operation of the attributes and thoughts through which the great dynamo operates or preceeds, he—an Ego—one of the expressions of infinity, will be enabled to free himself from the seeming environments of matter, and thus realizing his power, will assert his dominion over all he has been an agent in creating. And he has indeed assisted in creating—manifesting—all that is. Being a thought, an outbreathing of universal spirit, he is co-eternal with it.

In material concept, we do not begin to realize the extent of our wisdom. When we awaken to Egos, or spirit, consciousness—knowledge that we are Egos that have bodies or temples, and not bodies that have Egos—we see the object or reason of all symbols or manifestation, and begin to realize our own power over all created things.

And in this Aquarian age, great changes in nature's laws will be speedily brought to pass, and great changes in the affairs of humanity will result. The laws of vibration will be mastered, and through their operation material manifestations will be shaped and moulded to man's will. It is only a matter of time when all the necessities of life will be produced directly from the elements of the air.

It is well known by chemists that all manner of fruits, grains and vegetables are produced directly from the elements in the air, and not from the soil. The earth, of course, serves as a negative pole and furnishes the mineral salts of lime, magnesium, iron, potassium, sodium and silica, which act as carriers of water, oil, sugar, albumen, etc., and are formed by a precipitation or condensation of principles in the air, and not from the soil. This is a fact abundantly proved. Mr. Berthelot, a scientist of France, Tesla, the Austrian wizard, and our own Edison have long held that food can be produced artificially by a synthetic process from its elements.

Some six or seven extracts, as well as coloring material, are now being manufactured in this manner. Madder is made almost exclusively by this process now.

Mr. Berthelot, at one time the French minister of foreign affairs, possesses fame apart altogether from his political efforts. In his special domain of chemical knowledge he ranks among the first of his contemporaries. Chemical synthesis—the science of artificially putting organized bodies together—may be said to owe its existence to him. The practical results expected to flow from his experiments and discoveries are enormous. Thus, sugar has recently been made in the laboratory from glycerine, which Professor Berthelot first made direct from synthetic alcohol. Commerce has now taken up the question; and an invention has recently been patented by which sugar is to be made upon a commercial scale, from two gases, at something like 1 cent per pound. M. Berthelot declares he has not the slightest doubt that sugar will eventually be manufactured on a large scale synthetically, and that the culture of sugar cane and beet root will be abandoned, because they have ceased to pay.

The chemical advantages promised by M. Berthelot to future generations are marvelous. He cites the case of alzarin, a compound whose synthetic manufacture by chemists has destroyed a great agricultural industry. It is the essential commercial principle of the madder root, which was once used in dyeing, wherever any dyeing was going on. The chemists have now succeeded in making pure indigo direct from its elements, and it will soon be a commercial product. Then the indigo field, like the madder fields, will be abandoned, industrial laboratories having usurped their place.

But these scientific wonders do not stop here. Tobacco, tea and coffee are to be made artificially; not only this, but there is substantial promise that such tobacco, such teas and such coffees as the world has never seen will be the outcome. Theobromine, the essential principle of cocoa, has been produced in the laboratory, thus synthetic chemistry is getting ready to furnish the three great non-alcoholic beverages now in general use.

Biochemists long ago advanced the theory that animal tissue is formed from the air inhaled, and not from food. The food, of course, serves its purpose; it acts as the negative pole, as does the earth to plant and vegetable life, and also furnishes the inorganic salts, the workers that carry on the chemistry of life, setting free magnetism, heat and electric forces by disintegration and fermentation of the organic portions of the food.

But air, in passing through the various avenues and complex structure of the human organism, changes, condenses, solidifies, until it is finally deposited as flesh and bone. From this established scientific truth, it appears that, by constructing a set of tubes, pumps, etc., resembling the circulatory system, as well as the lung cells of the human mechanism, which is a chemical laboratory, where the chemistry of spirit is ever at work, changing the one essence of spirit substance to blood, flesh and bone, air may be changed into an albuminous pabulum, which may be again changed into the special kind of food required by adding the proper flavor, which may also be produced direct from the air.

There does not seem to be any reason why this substance, the basis of all food or vegetable growth, cannot, by drying and proper process, be made into material for clothing. Wool, cotton, flax, silks, etc., are all produced from the universal elements through the slow, laborious and costly process of animal or vegetable growth. Why not produce them directly?

Those who believe in a time of peace on earth, a millennial reign, certainly do not think that our present mode of producing food will continue during that age. Slaughter of animals, and fruit, grain or vegetable raising, leave small time for men and women to enjoy a condition foretold by all the seers and prophets. But under the new way of producing food and clothing, the millennium is possible.

And thus will the problem of subsistence be solved. No more monopoly of nature's bounties. An exchange of service will be the coin of the world instead of certain metals difficult to obtain.

A realization of this vision, or theory, that will for awhile be called visionary by most people, will mean Eden restored. Many people have wondered why, during the last few years, fruit pests have multiplied so alarmingly, and why cows are almost universally diseased and so much attention given to meat, milk, and butter products by Boards of Health, etc. There is surely a reason for all this. The One Life, Supreme Intelligence, or Divine Wisdom, that holds the worlds of space in their appointed orbits, surely knows all about the affairs of earth. When a new dispensation is about to be ushered in, old things begin to pass away.

All labor of preparing food and clothing, as now carried on, will cease, and the people, in governmental or collective capacity, will manufacture and distribute all manner of food and clothing free. Machinery for the production of everything necessary for man's material wants will be simple and easily manipulated. One-twentieth of the able-bodied population, working one or two hours a day, and shifting every week, or day, for that matter, with others, will produce an abundant supply. Neither droughts nor floods nor winter's snow can affect the supply. It can be made in Klondyke or the Tropics. Garments may be worn for a few days and then burned, and laundry work cease. Cooking will be reduced to a minimum, as the food will only need flavoring. No preparing vegetables, fruits, or cracking nuts; no making butter, or preserving meats. Men will not have to devote their lives to the endless grind of food production, nor woman to cooking, dish-washing, sewing, and laundry-work. Garments of beautiful design and finest texture will be made by machines invented for the purpose, ready for wear.

A dream, you say? I cannot admit that, in the face of the indisputable evidence it has already been able to produce, but what, if it were, at present, but the dream it may appear to be to the one who hears of its methods of operation for the first time, it is certain of future fulfillment. Do dreams ever come true? Yea, verily! All concrete facts are materialized dreams.

An Egyptian King dreamed, and the Pyramids of Cheops mass and miracle his vision. The Pyramids are encyclopedic of physical science and astral lore. The science of numbers, weights, measures, geometry, astronomy, astrology, and all the deeper mysteries of the human body and soul are symboled in these incomparable monuments.

A dream of an ancient alchemist solidified in stone, and the awful sphinx sat down in Egypt's sand to gaze into eternity.

Columbus dreamed, and a white-sailed ship turned its prow west and west. On uncharted seas, with an eternity of water ahead, he remembered his dream, and answered "Sail on!" to the discouraged mate, until he landed on the unknown shores of a most wonderful new world.

Michael Angelo dreamed a thousand dreams and sleeping marble awoke and smiled. Hudson and Fulton dreamed, and steamboats "run over and under the seas."

The Pilgrim Fathers dreamed and America, the "marvel of nations," banners the skies with the stars and stripes. Marcus Whitman and Lewis and Clarke dreamed long and hard and the bones of oxen and men and women and babies made a bridge over the desert sands and the mountain gorges to the shores of the Sundown Sea, and now the Pullman cars come safely over. Morse and Marconi and Edison dreamed strange wild dreams and concentrated intelligence springs from carbon-crucible and says to earth's boundaries, "Lo; here am I."

Vibration of etheric substance
  Causing light through regions of space,
A girdle of something enfolding
  And binding together the race—
And words without wires transmitted,
  Aerial-winged, spirit-sandalled and shod:
Some call it electricity,
  And others call it God!

A mechanic dreamed, and sprang upon his automobile, and drove it till the axles blazed and the spaces shriveled behind him. Men of high strung airy brains dreamed wondrous dreams, and now the eagle's highway and the open road of men lie parallel.

A musician dreamed a sweet, harmonious dream, and forth from a throat of brass directed by a million tiny fingers of steel, came the entrancing notes that have run riot through the singer's brain.

So let us dream on, men and women, of the day of rest that is already dawning in the heavens. No wonder that Paul said, "Now, brethren, are we the Sons of God, but it doth not yet appear what we shall be,"—as such. The morning light of that glad day now purples the mountains of faith and hope with its rays of glory.

And when Man is once fully alive to his own heritage, realizing the wonders and possibilities of his own body, and the power of his spirit to control it, and to provide for its needs, he will assert the divine right within him to be a spirit in command of its own temple, and the environment of that temple, and will rejoice in the revealed truth of his own divinity that alone can make him free.

TRANSMUTATION, OR TURNING WATER INTO WINE

THE Lord of transmutation has ascended the throne of Aquarius to rule the world for 2000 years.

Aquarius the fifth Son (Sun) of Jacob, circle, or to follow after, is Dan, Hebrew for Judge. Thus the day or time of judgment or understanding will have for its executor the revolutionary Planet Uranus, or as it is in Greek "Oranous." Uranus virtually means *Son of Heaven.*

This god is surely a suitable ruler for the Zodiacal Sign Aquarius the *Man.* "And then shall appear the *Sign* of the Son (Sun) of man in the Heavens."

The Solar System now being in Aquarius we may expect, and as a matter of fact, are experiencing the prophecies of great Astrologians as recorded in Matt. 24th, also Luke 21st.

In the Judgment Day, or time of knowledge, we are due to realize the process by which base metals are transmuted into gold.

The word gold comes from Or, a product of the Sun's rays, or the breath of life.

Life or Spirit breathed into man precipitates brain cells and gray matter which creates or builds the fluids and structure of physical man.

Or, is the seed or the W*or*d—L*or*d, etc.

"In the beginning was the w*or*d . . . the w*or*d was God."

Even our great Theologians have admitted that G o d means Power.

Thus the emanations from Sun, basic material, are changed to gold and the process, eternally proceeding, is being recognized by man at the present day, due to the fact that the Planet of gold, *Or*anous, is now ruling Earth and thereby bringing good judgment upon the people, and the thrones and scepters of rulers lie scattered and crushed along the highway of Nations.

Both in Greek and Hebrew any fluid, air or ether was called water until organized; then it was wine. The rain that falls on the ground and is taken up into the organism of tree, vegetable or fruit is changed into wine, i. e., sap or juice.

The parable of turning water into wine at the marriage at Cana in Galilee is a literal statement of a process taking place every heart beat in the human organism.

Galilee means a circle of water or fluid—the circulatory system. Cana means a dividing place, the lungs or reeds, the tissue and cells of the lungs.

Biochemists have shown that food does not form the organic part of blood, but simply furnishes the mineral base by setting free the inorganic or cell-salts contained in all food-stuff. The organic part, oil, fibrin, albumen, etc., contained in food is burned or digested in the stomach and intestinal tract to furnish motive power to operate the human machine and draw air into lungs, or Cana, thence into the arteries, i. e., air carriers. Therefore, it is clearly shown that air (spirit) unites with the minerals and forms blood, proving that the oil, albumen, etc., found in blood, is created every breath at the "marriage in Cana of Galilee."

Air was called water or the pure sea, viz.: Virgin Mar-y. So we see how water is changed into wine—blood—every moment.

In the new age, we will need perfect bodies to correspond with higher vibration, or motion of the new blood, for "old bottles (bodies), cannot contain the new wine."

Another allegorical statement typifying the same truth reads, "And I saw a new Heaven and a New Earth," i. e., new mind and new body.

Biochemistry may well say with Walt Whitman, "To the sick lying on their backs I bring help, and to the strong upright man I bring much needed help." To be grouchy, cross, irritable, despondent or easily discouraged, is prima facie evidence that the fluids of the stomach, liver and brain are not vibrating at normal rate, the rate that results in equilibrium or health. Health cannot be qualified, i. e., poor health or good health. There must be either health or dis-health; ease or disease. We do not say poor ease or good ease. We say ease or dis-ease, viz., not at ease.

A sufficient amount of the cell-salts of the body properly combined and taken as food — not simply to cure some ache, pain or exudation—forms blood that materializes in healthy fluids, flesh and bone tissue.

We should take the tissue cell-salts as one uses health foods, not simply to change, not health to health, but to keep the rate of blood vibration in the tone of health all the time.

Biochemistry is the sign board pointing to the open country, to hills and green fields of health and the truth that shall set the seeking Ego free from poverty and disease. Conservation and transmutation obtains in all the commercial world. The force of falling water is transmuted into the product of the factories. Steam, the vibration of copper and carbon discs that turn night into day, the "chariots that run like lightning and jostle each other in the streets" are the effects of the transmutation of base or basic material.

On some fair tomorrow when the subtle vibrations of the Aquarian Age, directed by *Or*anous shall have awakened and called to action the millions of dormant cells of the wondrous brain, man will, by the power of the lost word, conserve and transmute the material substance of his body, the soul, I O H N or J O H N, and with the "product," the precious ointment—*oil*—triumph over the cross at Golgotha and ascend to the pineal gland that transmits the christed Son to the Optic Thalmus, the all-seeing Eye of the chamber and thus furnish "light to all that are in the house."

In these latter days our business world has been dominated by a great oil trust, and petroleum, mineral oil, petra, stone, rock or mineral and oleum-oil was exploited by Rockefeller, a mineral fellow, and then by the laws of transmutation changed into gasoline.

The transmutation of gasoline by the miracle of the "conservation of energy" causes the "Ascension" of the air ship, and the pathway of the Eagle and the open road of man lie parallel across the vaulted sky.

And when the Ego shall have triumphed over the carnal mind and transmuted the crude soul fluids into the gold of the "New Wine" it will ascend to the Father, the upper brain.

"And the temple needs no light of the Sun by day nor moon by night, for the light of the Lord doth lighten it."

Thus will the "Blood of Christ cleanse from all sin."

Our Lord Oranous, Son of Heaven, ruling the Sign of the Son of man, may then exclaim, "I have fought the good fight, I have finished my course."

PSYCHOMETRY

THE word psychometry was coined by the noted scientist, Joseph Rhodes Buchanan. It is derived from psycho, or Psuche, the Greek for soul, and mitron, meter, to measure. Therefore, the word psychometry means literally *soul measurement.* Soul is from the Hebrew Nephesh, meaning breath, and the Greek word Psuche,—breath.

"God breathed into man the *breath* of life and he became a living *soul.*"

Thus every atom of the body is Soul Stuff and the complete body is the soul that can be *saved* or *lost.* While the entire body is soul, the scriptures refer, in most instances to the nerve fluids, gray matter of the brain and marrow of the spinal cord (River of Jordan) as soul.

The word breath is translated soul more than 500 times in the bible. Paul writes of body, soul *and* spirit.

The contention that psychometry is a science, and therefore worthy of recognition, is sustained by two indisputable facts. First, chemists all agree that there is but one substance in the universe, which in different degrees of fineness, attenuation or rates of motion, causes all appearances, all forms, all that we see, hear, feel, smell, or in any way sense, thus corroborating Herbert Spencer's statement that "all things proceed from one universal energy."

Second fact: Every person who is at all sensitive to impressions is more or less affected by the influence, the aura, of clothing, of works of art, or any product of man's hands, as well as natural products, but more especially are they affected by reading the written or spoken words of individuals.

No theory seems better established as a fact, than the theory that the particular vibration, or degree of intensity of thought, or the particular quality of a writer or speaker, is transmitted to the reader or listener.

We are not yet able to tell by what subtle, wireless telegraphy, chemical affinity, or alchemical transmutation, this marvelous operation is performed; but we must recognize the fact.

Chemistry is now merging into alchemy, from which it originally sprung. Modern chemistry is but the material side or effect of alchemy. All real scientists and especially the advanced chemists agree that the universal substance in its higher rate of motion, constitutes the higher intelligence, spirit or energy.

Other or lower rates of activity manifest in the gases, water, vegetable, wood, stone, etc., etc.

Thus, ice is crystallized or frozen water while water is itself a product of rates of motion called oxygen and hydrogen—two of hydrogen to one of oxygen.

But water disappears when the two gases are set free or separated.

The new chemistry has discovered the truth that all gases, all so-called elements, can be split, chemically, and resolved into the unknown and unamenable Absolute.

The spiritual scientist does not recognize law as a separate agent from himself, but being "led by the spirit" he fully realizes that he is himself a phase of the law in operation.

By thus recognizing the universal co-operation of the attributes or thoughts through which the great Dynamis operates or proceeds, you, a spiritual entity, one of the thoughts or words of God, are enabled to free yourself from the seeming environments of matter and thus realize your dominion over all you have been an agent in creating.

And you have assisted in creating—manifesting—all that is.

Being a thought, an outbreathing of universal spirit, you are co-eternal with it. The substance everywhere, omnipresent, was of course never created.

These thought-vibrations are materialized, analyzed, understood and described as a painter explains the effects, lights and shades of his pictures, or a mechanic the results of his handiwork.

In material concept we do not realize the extent of our wisdom. When we awaken to spirit, consciousness—knowledge that we are Egos that have bodies or temples and not bodies that have Egos—we see the object or reason of all symbols or manifestations, or formed things, and we spell the words again and this is called psychometry.

With a full understanding of the Oneness of life, and that all life is but a phase of Eternal Life, comes the power to psychometrize. If we can cognize the great truth that the products of nature are likewise the results of our own intelligence, we begin to see why it is possible to read the history of a certain article by coming in touch with its vibrations.

Let us make the proper distinction between reason and intuition. Intuition is knowing. Reason is only a supposition, that a certain thing or theory is true. If it be demonstrated that the theory is false, the structure or argument built upon the sand-formation of reason, falls. Intuition does not depend upon reason or so-called logical deductions.

To psychometrize we should wait for "the still small voice" of intuition, and not attempt to find truth by the material roads of reason and logic.

Relax all bodily functions and thus quiet material operation and you may read so-called inert matter, as you would read printed pages. Psychometry is miracle made natural, or occultism engrafted into everyday life.

"Nothing is hidden that shall not be revealed," and humanity stands at the door of the New Day when "none shall say unto his neighbor, "Know the Lord," for all shall know Truth and Truth shall set them free."

IT:
THE ETERNITY OF PERFECTION

A CHILD brought to its mother a piece of ice, and asked: "What is this?"

The mother answered, "it is ice."

Again the child asked, "What is there in ice?"

The child desired to find the water in the ice, and it procured a hammer, and pounded the piece of ice into little bits and the warm air soon changed all the ice to water. The child was grievously disappointed, for the ice that the child supposed contained water had disappeared.

And the child said, "Where is the ice that contained this water?"

And so it came to pass that the mother was compelled, by the child's persistent questions, to say, "ice is all water; there is no such thing as ice; that which we call ice is crystalized or frozen water."

The child understood.

A student brought to his teacher some water and asked, "What is water? What does it contain?"

The teacher answered, "Water contains oxygen and hydrogen," and then explained how the two gases might be separated and set free by heat.

The student boiled the water until all of the molecules of oxygen and hydrogen had been set free, but he was surprised to find that all of the water had disappeared.

Then the student asked of the teacher, "Where is the water that held the gases that have escaped?"

Then was the teacher compelled by the student's persistent questions to answer, "Water itself is the product of oxygen and hydrogen. Water does not contain anything other than these gases. In reality, there is no such substance or fluid as water; that which we name water is a rate of motion set in operation by the union of two parts of hydrogen with one part of oxygen, and of course the phenomenon disappears when the union of the gases is broken."

The student understood.

A devout scientist presented himself before God and said, "Lord, what are these gases men call oxygen and hydrogen?"

The good Lord answered and said, "They are molecules in the blood and body of the universe."

Then spake the scientist, "Lord, wilt thou tell me of the kind of molecules that compose Thy blood and body?"

The Lord replied, "These same molecules, gases or principles, compose my blood and body; for I and the universe are one and the same."

Once again the scientist said, "My Lord, may I ask, then, what is spirit and what is matter?"

And thus answered the Lord:

"As ice and water are one, and the gases and water are one, so is spirit and matter one. The different phases and manifestations cognized by man in the molecules of My body—that is, the universe,—are caused by the Word; thus, they are My thoughts clothed with form."

Now the scientist felt bold, being redeemed from fear, and asked "is my blood, then, identical with Thy blood in composition and Divine Essence?"

And the Lord said, "Yea, thou art one with the Father."

The scientist now understood and said,

"Now mine eyes are opened, and I perceive that when I eat, I partake of Thy body; when I drink, I drink of Thy blood; and when I breathe, I breathe Thy spirit."

So-called matter is Pure Intelligence and nothing else—because there is not anything else.

Pure intelligence cannot progress or becomes better. There is nothing but Intelligence. Omnipresence, Omnipotence, Omniscience must mean Intelligence; therefore these terms are all included in the word.

Let us adopt a short word that will express all that the above written words are intended to express, namely, the word IT. "I" stands for all—the eternal I. "T" stands for operation, manifestation, vibration, action or motion. The "I" in motion is "T," or Crossification, viz., the T-cross. We say, "IT" rains!" "IT is cold!" "IT is all right!" What do we mean by "It?" Who knows? Some say, "The weather!" Others, "Natural phenomena!" Very well, then, —what do we mean by "the weather," or "natural phenomena"? Why, just It, of course!

IT does not progress; it does not need to. IT forever manifests, operates, differentiates and presents different aspects or viewpoints of ITSELF. But these different phases are neither good, better nor best, neither bad nor worse—simply different shades and colorings of the One and Only Intelligence.

Every so-called thing, whether it be animal, vegetable, or mineral, molecule or atom, ion or electron, is the result of the One Intelligence expressing itself in different rates of motion. Then what is Spirit?

Spirit means breath or life. Spirit, that which is breathed into man, must be intelligence, or man would not be intelligent. Non-intelligent substance, which is, of course, unthinkable, would not breathe into anything, nor make it intelligent if it did. Therefore we see that Spirit, Intelligence and Matter are one and the same Esse in different rates of motion.

So-called molecules, atoms, electrons, know what to do. They know where and how to cohere, unite, and operate to form a leaf or a flower. They know how to separate and disintegrate that same leaf or flower. These particles of omnipresent life build planets, suns, and systems; they hurl the comet on its way across measureless deserts of star-dust, and emboss its burning path.

From the materialistic and individual concept of life and its operations, it is pitiable and pathetic to view the wrecks along the shores of science. It is only when we view these apparently sad failures from the firm foothold of the unity of being and the operation of wisdom that we clearly see in these frictions and warring elements and temporary defeats and victories the chemical operation of Eternal Spirit—operating with its *own substance*—its very self. It is only through the fires of transmutation that we are enabled to see that all life is one Eternal Life and therefore cannot be taken, injured, or destroyed.

The fitful, varying, changing beliefs of men in the transition stage from the sleep and dreams of materialism to the realization of the Oneness of Spirit show forth in a babel of words and theories, a few of which I shall briefly consider, beginning with the yet popular belief in Evolution:

The evolutionary concept has its starting point in the idea (a) that matter—so-called—is a something separate from mind, intelligence, or Spirit; (b) that this matter had a beginning; (c) that it contains within itself the desire to progress or improve; and finally, that the race is progressing, becoming wiser, better, etc.

Against this assumption, I submit the proposition that the Universe,—one verse,—always existed without beginning or ending, and is and always has been absolutely perfect in all its varied manifestations and operations.

A machine is no stronger than its weakest part. If the self-existing universe is weak or imperfect in any part, it must, of necessity, always have been so. Having all the knowledge there is—*being* all,—it is unthinkable that there is any imperfection anywhere. Everything we see, feel, or taste, or in any manner sense, is perfect substance, condensed or manifested from perfect elements, but all differ in their notes, vibrations, or modes or rates of motion. A serpent is as perfect, therefore as good, as a man. Without feet, it outruns a man; without hands, it outclimbs the ape, and has been a symbol of wisdom through all the ages. Man is an evil thing to the serpent's consciousness. Neither are evil—nor good. They are different expressions or variations of the "Play of the Infinite Will."

The brain of the jelly-fish is composed of the same elements, of the same substance as the brain of a man, merely of a different combination. Can man tell what the jelly-fish is thinking, or why it moves and manifests its energy thus or so? How, then, is man wiser than the jelly-fish because his thoughts are of a different nature, and operate to different ends?

Wisdom—all there is—simply operates, manifests, expresses forms, or creates with, or of, self-existing substance. As wisdom is without beginning or end, so are all its operations or manifestations without beginning or end. If the race is constantly evolving to higher standards and loftier conceptions, why send young men and women to Rome and Florence to study the "Old Masters?" If man has evolved up from the "lower forms of life," why has he spent so much time, money, and brain energy in trying to understand these lower forms, and to do what they do? Why does he not remember and retain the power of his earlier states of manifestation?

The eagle must wonder, as it watches man's efforts and failures to perfect his flying machine, how long it will be before he evolves up to the science of the birds, i. e., the science of flying. Modern man is now taking his first lessons in condensing or materializing air, while through unnumbered ages the spider has performed the miracle without the necessity of first attending a school of chemistry. The *modus operandi* by which the spider forms his web from air is the despair of science. The wisdom of the ant or beaver strikes dumb all the believers in the Darwinian dream. The perfect co-operative commonwealth of the bees is still the unattained ideal of man.

Beneath the soil upon which falls the shadow of the throne of Menelik, the Abyssinian King, are layers and strata of buried civilizations, and astronomers in China mapped the Heavens, named the stars, calculated eclipses and the return of comets ages before Moses led the Hebrews out of bondage, or the walls of Baalbeck cast a shade for the Arab and his camel.

The evidences and witnesses of the wisdom of men on earth hundreds of thousands of years ago confront the scientific investigator at every turn. Here the Rossetta Stone, and there the Inscribed Cylinder of Arioch or Statue of Gudea, King of Chaldea. Prophecies, inscribed on Cunieform tablets of Clay, foretelling the building of the Pyramids, are brought to light by the excavator; and the history of the Chinese Empire, running back in links of an unbroken chain for one hundred and fifty thousand years, forever refute the theory of the "Descent of Man!" Side by side with the ancient Asiatics who knew all that we today know, dwelt the Crystal, the Cell, the Jelly-Fish, the Saurian, the Ape, and the Cave-Man. Side by side with the masons who could build arches of stone in ancient Yucatan that mock at the ravages of Time, lived and wrought the ant, operating in its co-operative commonwealth of which man can still only dream. Side by side with the cave men and cannibals dwells the spider, whose operation in aerial elements is the despair of chemical investigators. And when Solomon's golden-spired temple illuminated the Holy City, or the tower of Babel grew toward the clouds, or the Mound Builders recorded their history in rock and soil, the eagle and the dove calmly floated in the air and wondered when men would evolve to their plane of science. They are wondering still.

Exponents of the evolutionary theory never tire in quoting Professor Huxley. One who has not read the writings of this eminent scientist would be led to believe by the statements of his followers that he had positive views on the great question of force and matter. Following is an extract from a letter written by Professor Huxley to Charles Kingsley, under date of May 22nd, 1863, taken from the published letters of Huxley by his son, Leonard:

"I don't know whether Matter is anything distinct from Force. I don't know that atoms are anything but pure myths—'Cogito ergo sum' is to my mind a ridiculous piece of bad logic, all I can say at any time being 'Cogito.' The Latin form I hold to be preferable to the English 'I think,' because the latter asserts the existence of an Ego—about which the bundle of phenomena at present addressing you knows nothing. I believe in Hamilton, Mansell, and Herbert Spenser, so long as they are destructive, and laugh at their beards as soon as they try to spin their own cobwebs.

"Is this basis of ignorance broad enough for you? If you, theologian, can find as firm footing as I, man of science, do on this foundation of minus naught—there will be naught to fear for our ever diverging. For you see, I am quite as ready to admit your doctrine that souls secrete bodies as I am the opposite one that bodies secrete souls—simply because I deny the possibility of obtaining any evidence as to the truth or falsehood of either hypothesis. My fundamental axiom of speculative philosophy is that materialism and spiritualism are opposite poles of the same absurdity—the absurdity of imagining that we know anything about either spirit or matter."

Huxley admitted that he did not know.

As the appetite craves new chemical combinations of food from day to day, so does mind crave new concepts of infinite life. The word "Infinite" defines an endless differentiation of concept.

If the Spiritual Consciousness,—the "mighty Angel" that the clairvoyant seer, John the Revelator, saw descending out of the Heavens, shall carry away the pillars of material evolution, a Temple of Truth divinely fair will spring, Phoenix-like, to take its place. Eyes shall then be opened, and ears unstopped. Man will then realize that the so-called lower forms of life are just as complex, wonderful, and difficult to form as the organism of man—that proto-plasm is as wonderful in any other form as in the gray matter of the human brain, which is only another form of its expression,—that the molecular composition of a jelly-fish puzzles the greatest chemist, and the wisdom of a beaver is enough to strike dumb all the believers in the Darwinian fairy tale.

And has the dream of good and evil any better foundation than has this one of material evolution? We are here to solve the problems of life, not to evade them; and to name the mighty operations of Eternal Wisdom good and evil is simply evading instead of solving.

The universal Principle, Spirit, or God, is impartial. Saint and Sinner are one in the Eternal Mind. God, or Infinite Life is not in the least injured by so-called good or evil. The Spiritual Ego is the interested party and must work out its *own* Salvation. There is no point in the universe better, higher, or nearer God, or the centre, than any other point. All places are necessary, and no one is favored over any other. As Huxley well said, "Good and evil are opposite poles of the same absurdity." Good must have evil for its opposite, if it exists at all. He who would realize Being must get rid of the concept of good, as well as the concept of evil. Good and evil are qualifications, and Being does not admit of qualification or grades. It simply is. The ideal we call good eternally exists, but its name is wisdom's operations. Nothing is low or high, good or bad, except to individual concept that allows comparison. "Comparisons are odious."

Physical Science, so-called, declares in its text-books that light travels from the sun to the earth in eight minutes,—a distance of about ninety-five million miles. To question this statement a few years ago meant ostracism from the circle of the elect who knew things. But today the iconoclast stands at the gate of the temples of learning and batters at the walls with the hammer of Thor. Fear and trembling seize upon the votaries of material gods as they see evolution, progression, the theories of electricity, light, and heat, good and evil, all cast into the crucible of truth for transmutation in the Divine Alchemy of Being, all dissolving as pieces of ice of different sizes and shapes change to water.

The present day chemist, as he begins to tread the soil where stood the ancient alchemist, tells us that light and heat are simply rates of motion of a substance that does not travel from star to star or from sun to planet, but vibrates in its place at rates directed by the Eternal Word. This substance, aerial or etheric, does not travel—it is everywhere present—the body of omnipresent being.

Men now dare assert that there is no evidence that the sun is hot, but that there is evidence that the sun is the dynamo of the Solar System and so vibrates the etheric substance that light, heat, cold and gravitation are produced,—not as entities separate from the universal elements, but as results or effects produced by different rates of motion of the molecules of the wire—molecular motion—or of the air or etheric substance, as in wireless telegraphy.

Another ancient belief, now obselete, is the progression of man in a better state of existence after death or cessation of bodily functions. This idea had its origin in the fallacy that there were grades of goodness in the Divine Mind, and that somehow we are not treated right during earth life, and that, in consequence, we must be rewarded by an easy berth "over there." But we now see quite clearly that the great cause of life and all its operations would be unjust to withhold from its sons and daughters for one moment anything that belonged to them. If the Cause ever does wrong, we see no reason why it should repent and do right. If the Cause ever failed in the least particular to give just dues, it may do so again at any time. The "better state of existence" mentioned above can only come through wisdom obtained here and now; thus will man "work out his own salvation."

The time was, and not so very long ago, when the recognized scientist believed that there were about seventy-four elements, indivisible, separate, and distinct; but the alchemical iconoclast with his hammer of truth has pulverized the fallacy and remorselessly hammered and pounded the seventy-four faces into one countenance.

For a long time, hydrogen gas, the negative pole of water, was supposed to be indivisible beyond all question; but the present day chemist knows it is only an expression of yet more subtle molecules back of which, "Standeth God within the shadow keeping watch above His own."

A post mortem examination of some of the wreck's along the shores of the troubled sea of science discloses a belief that the Ego is an individual, who through knowledge of its divine origin may draw unto itself all things it may desire! But as fast as the Sleepers awaken they see that each Ego is only "part of one stupendous whole" that does not draw unto itself anything. That there is no law of attraction for the eternal substance is everywhere present and each one uses exactly that portion prepared for him from everlasting unto everlasting.

When the continuity of life was first demonstrated beyond question those who caught the first dispatches from disincarnate spirits sprang forth from their beds of material sleep and with half-opened eyes only saw the great truth through "a glass darkly." Then came a babel of words. They jabbered a jargon that needed translation to be understood. The ideas of progression in earth life that obtained among men was transplanted to the spirit realm and we were told by the votaries of spiritual philosophy that men and women had great opportunities for progression after leaving the flesh.

As the idea of a commencement of the universe was a common belief among those asleep in material consciousness, being the corner stone of evolution, so the idea obtained that the individual had a commencement in the maternal human laboratory. As these half-awakened individuals could not comprehend that an action contrary to their concept of good could possibly be caused by Infinite Intelligence they concluded that the so-called bad actions of men and women were prompted by evil earth-bound spirits. These people—many of them—also thought that the main object of the existence of Spirits in the Spirit realm was to gather information about mines, and stocks, and bonds, and lotteries, and races and thus assist poor mortals to get-rich-quick. It was supposed that these spirits were posted in regard to deeds and wills and knew when wealthy relatives would shuffle off the mortal coil or when undesirable wives or husbands would "pass out."

But at last the sun of Truth pierced the darkness and the jargon of selfishness changed to the "New Song." We now clearly see that each spirit is a part or attribute of the One Eternal Spirit—therefore has existed always and that the process of generation deals with flesh clothing, or mask for the spirit in which it performs a necessary part in the creative process. The word "person" is derived from a Greek word, *Persona,* meaning mask.

We see that the phenomena we have called obsession by evil spirits is God's surgery or dynamic operation in his own temple quite as impossible for us to understand in our present environment as it is for the child to understand the wisdom and necessity in the operation of the adept surgeon.

And, finally, we now see and realize fully that Eternal Wisdom without beginning or end of days does not progress before entering a temple of flesh, while it occupies it, or after it leaves it. All creative or formative processes may properly be termed operations of wisdom or Eternal Life.

In the unwalled temple of the Now, beneath its roofless dome there is no progression, but a constantly moving panorama forever presenting to consciousness new phases of the absolute.

The men and women who do things take hold of opportunities and material that they find all about them now, and operate with them, astonishing results following the efforts of all who recognize that eternal force has use for them NOW to carry out the divine plan. We are all operators or workmen in the divine workshop, and the Divine Intelligence, the eternal IT, made no mistake in placing any of us here, but does insist that we recognize that NOW is the time and Here is the place to do our best. As the Great Cause does not need to first practice on lower forms in order at some future time to attain perfection, we must recognize and practice *being* in the present, instead of *becoming* in the future, for the Eternal Now is all the time there is.

"But," you say, "your science has taken away my God, and I know not where you have lain Him." On the contrary, I have brought you to the one true God, "which was, and is, and evermore shall be."

The fifth verse of the last chapter of the book of Job reads as follows:

"I have heard of thee by the hearing of the ear; but now mine eye seeth thee."

The wonderful writings and scientific statements found in that Book of all books, the Christian Bible, were recorded at dates covering thousands of years by men and women who never heard of each other. Some of these teachers lived away back in the age when the solar system was swinging through the zodiacal sign, Taurus; when Phallic worship prevailed; when the number six was understood as sex, and the creative or formative principle operating through the sex functions was worshipped as the very Holy of Holies. Other teachers who contributed to the knowledge of life and its operations contained in the Biible, lived in the age of Aries, a fire sign, when fire and sun were worshipped as the essence of God; and as heat, the cause of the phenomenon called fire, cannot be seen, it was a reasonable thing to say that "no one can see God and live." So then, it depends upon the point of view one has of God, or the spirit of things, whether he says, "No one can see God and live," or says, "Now mine *eye* seeth thee!"

I think the writer of the book named Job must have lived

more than eight thousand years ago, even before the Taurian age—symbolized by the Winged Bull of Nineveh—which was in the Gemini age, the age of perception and expression, being an air age. Let it be understood that an age in this connection means twenty-two hundred years, the period for the solar system to pass across one of the signs of the zodiac. In an air age, Egos awaken to their divine heritage, and realize their Godhood. The writer of Job, then, living in the Gemini or air age, could see God and live. Our solar system has entered the sign Aquarius, another air sign, and the spiritualized elements so act upon our brain-cells that we are able to understand the teachers of a past air age, and also see God and live.

Carlyle, the prince of literary critics, said "The book of Job is the most wonderful and beautiful literary production ever given to the world." Certainly the scientific truths of astrology and alchemy, and of the Spirit's operation in flesh, as set forth in that book, are without a parallel. The letters J, O, B, have an occult, scientific meaning, I and J are the same IOB meaning the same as JOB. I means the Eternal I. All the Hebrew letters were formed from I. O means the universe, without beginning or end, and B means Beth, a body, house, church, or temple. Therefore, GOD, or all, may be discovered as seen in JOB or IOB. The word, Job, has no reference to a person. The name, or letters of the word, symbolize principle, the same as wisdom, knowledge, intelligence, or Christ, or Buddha. We symbolize the principles of our government in personalities, and picture them in the form of a man or woman, namely, Uncle Sam, or Columbia. But we do more than that: we put words in their mouth, and make them utter speech. And shall we ignore these facts when dealing with the record of past ages? One record plainly states that Jesus spake only in parables.

But let us consider more closely the discovery of God. The numerical value of G.O.D., according to ancient Kabala, is nine—the all of mathematics—no person is alluded to. If the statement, "I and the Father are one," is true, the "I" must be the Father manifested or expressed. As it is not possible to conceive of the Father except through expression, we must conclude that manifestation in some form of so-called matter is eternal—the great necessity—and has therefore always been.

It is quite reasonable to think that some oxygen and hydrogen has eternally existed in gaseous form, some in the combination that causes water and some in the concrete or concentrated form known as ice. Then upon the postulate that Spirit and matter — that is, bodily or material expression —are one, it follows logically that matter, including the physical body or temple of man, is as necessary to the Father-Mother principle while held in a given rate of activity or expression as this life essence is necessary to matter, or the physical structure of man. I see oxygen and hydrogen when I look at the manifestation we call ice. When I see water, I know just how oxygen and hydrogen appear when united. So when I look at any form of so-called matter, I know exactly how God appears at that particular time and place. I do not see the effect or works of God, but I see *God,* and just as much of God, face to face, as I am capable of seeing or recognizing at a certain time.

Step by step, the scientific investigator is being led to the threshold of the awful, absolute Truth, that all matter, or substance, or energy or force,—call it what you may—is not only intelligent, but is *Pure Intelligence* itself. Atoms, molecules, electrons are but expressions of rates of motion of pure Mind, Thought, or Intelligence that man has personified and called God. Ice is not permeated with water, or controlled by water. Ice *is* water. Matter is not controlled by mind; mind and matter are *one.* A high vibration of mind does control, to a certain extent, a lower vibration of mind, as water may carry a lump of ice here or there, water being a more positive rate of activity of the same thing. The particles, so-called, of matter know what to do. The atoms that compose a leaf know when to cohere and materialize a leaf, and they know how and when to disintegrate and dematerialize it: "Thou shalt have no others gods."

I hold in my hand that particular form of the one thing called a rose. Material thought says it is made by God, or that God is in the rose or back of it, or that God caused or created it; but when Spirit, the I Am, asks where is the God that created the rose, where has he betaken Himself, material belief is silent. But hold a moment! I have here a bud, a half-formed rose. If God makes a rose, He must continue the work to completion. Ah, speak softly! Look closely! The rose is now being made, and you say God is making it. Yes, you said God made this full-blown rose. Well, then, He is surely now at work on this half-blown rose. Bring on your spectroscope, your microscope! Quick, now, you chemist! Bring on your test-tubes, your acids, and alkalis, your spectroscope and X-ray. Analyze, illuminate, and magnify! Now we shall discover God. He is here at work before our eyes.

What do you see, chemist? What do you see, scientist? Ah! I know what you see. My experience in the realm of matter and of Spirit tell me what you see. O thou stupendous sex force—sex—days of creation, thou Father-Mother Yahveh, thou divine male and female, thou eternal positive and negative dynamis! We now behold thee operating. Out from the chemicalizing mass of God's creative compounds, out of the quivering, vibrating substance, slowly comes forth the rose. But are you sure it is a rose? Hold a moment. What is a rose? Of what material is it formed? Ah! the chemist speaks—he of the crucibles and test-tubes and acids! Hear the chemist!

He says, "The rose is made from the universal substance," or "The rose is universal substance, in a certain rate of activity." Thanks! Blessed be the chemist! Universal—one verse—one substance—no other substance—God is the rose, or the smile we call a rose—God is again manifested in the great Eternal IT, for which there is no other name.

> "Acids and alkalis acting,
>   Proceeding and acting again,
> Operating, transmuting, fomenting,
>   In throes and spasms of pain—
> Uniting, reacting, atoning,
>   Like souls passing under the rod—
> Some people call it Chemistry,
>   Others call it GOD."

Job did not say, "I see the thoughts of God," nor did he say, "I can fathom the mind of God." The plan cannot be seen; but that which is planned—a planet—can be seen. One may see the substance of God without understanding the mind of God.

Let us hear Emerson on this stupendous, glorious theme:

> "The great idea baffles wit;
> Language falters under it;
> It leaves the learned in the lurch—
> Nor art, nor power, nor toil can find
> The measure of the Eternal Mind,
> Nor hymn, nor prayer, nor church."

O thou ever-present Divine Mind and Substance! We now fully realize our oneness with thee, and bathe and revel in thy glory. The mighty Angel of Reality has torn the veil of illusion, and we see the celestial City of Truth with wide-open gates and the white light of Eternal Love forever upon its streets.

O thou, in the shadow of sickness and trial, "Take up thy bed and walk; thy sins be forgiven thee."

ALCHEMY, BIOCHEMISTRY, WIRELESS TELEGRAPHY AND MENTAL HEALING

Behind chemical phenomena,
"Standeth God within the shadow,
Keeping watch above His own."

BIOLOGISTS and physiologists have searched long and patiently for a solution of the mystery of the differentiation of material forms.

The writer does not believe there is any such thing as an element different from the universal substance in its last analysis, but that so-called "elements" are different rates of motion, or vibration of *one substance.*

No ordinary test can detect any difference in the ovum of fish, reptile, animal, bird, or man. The same mineral salts, the same kind of oil, albumen, fibrin, sugar and carbon is found, not only in the egg or germ of all forms of life, but in the substance or tissues of the bodies of all the varied expressions of materiality.

The answer to this "Riddle of the Sphinx" is found where Bio, or life, Chemistry merges into alchemy, over the door of which is written, "It is finished"—"Let there be light."

Professor Loeb says: "The ultimate source of living matter is mineral." To the Biochemist the above is a truism. There is no such thing as inert, or dead, matter, All is life.

The base of manifestation is mineral. Out of the dust, ashes, or minerals of the earth, physical man is made.

The twelve mineral salts are the basis of every visible form, animal or vegetable. No two different forms have exactly the same combination of the minerals, but all have the same minerals. These minerals, inorganic salts, are the twelve gates of precious stones described by John in his vision.

When the Divine Word speaks the mineral atoms, or

molecules, of its body into a certain formula or combination, a germ or egg, which is the basis or nucleus of the form to be manifested, materializes. This little plexus of intelligent atoms then commences to attract to its centre by the law of chemical affinity, which is only another way of saying God in action, other atoms known as oxygen, hydrogen, nitrogen, etc., and thus materializes them, until the building is completed according to the plan of the architect or designer. Thus the Word, operating through chemistry, is the Alpha and Omega.

There would be no eagle, fish, horse, or man without the Word, Divine Wisdom, and there would certainly be no Word if there was no substance to obey the Word, and likewise there would be no substance if there was no law of chemical affinity, or action and reaction, whereby the operation of materialization and dematerialization may be carried on.

It will be demonstrated in the near future that so-called nitrogen is mineral in solution, or ultimate potency, which explains the reason why nitrogen enriches the soil.

The Atmospheric Product Company at Niagara Falls, whose promoters extract and precipitate nitrogen from the aerial elements by electrical process, are the forerunners of machines that will manufacture food and clothing direct from air, and also produce heat or cold as needed by different rates of vibration of the substance, body of God, everywhere present. The Boston American's leading editorial in an issue the latter part of May, 1918, said:

"According to Mr. Kebler, who is one of the Government's good workers in the field of science, there are, in the air that surrounds our planet, about 400 TRILLION tons of nitrogen.

That is a good deal of nitrogen. Multiply the 400 trillion tons by a hundred dollars, which would be a cheap price for a ton, and you see that much wealth is floating around in the air.

Nitrogen is extracted from the air by the use of electricity, and all intelligent Governments are at work getting it out.

There is plenty for all. You cannot exhaust the supply of

nitrogen in the air—as you might exhaust coal in the beds, or oil in the wells.

After the nitrogen is used by human beings, animals or plants, it is released and goes back into the atmosphere again.

And you cannot exhaust, either, the power that produces the electricity that takes nitrogen out of the air and gives it back to man for his fertilizing servant.

The electric power is supplied by waterfalls. After they have gone down over the cataracts and created their electric force, the sun lifts them up into the air again in clouds from the ocean, and they fall down and do the work over again.

It is as good an imitation of perpetual motion as you could care to have."

The St. Louis Post-Dispatch of August 31, 1902, contained an article under headlines as follows: "How Food May Be Made From the Air We Breathe," "The Latest Marvel of Science," "Atmospheric Food Company Incorporated at Niagara Falls."

"The interesting article in the Sunday Post-Dispatch, describing how modern invention will enable us to draw new food supplies from the air itself showed how an up-to-date newspaper keeps its readers informed on the latest scientific discoveries.

Readers of this article know that science can now produce nitric acid by burning the air by means of powerful electric arcs. The acid is made cheaper than it has heretofore been obtained by chemical action. And this cheap production means cheap nitrates. Cheap nitrates means more bushels of wheat per acre, and an increase in all kinds of nitrates. Now we can obtain these fertilizers anywhere, by setting an electric current to work. The possibilities are stupendous."

This gas nitrogen, which makes four-fifths of the atmosphere, was a puzzle to early science. It seemed dead and useless. It was looked upon merely as a filler, to prevent the more active oxygen from burning the earth up. It appears that it may yet become an exhaustless common bank, from which humanity may draw wealth for all time. The story of its exploitation is one of the fairy tales of science.

By chemistry, the court of last resort, will man come into

his divine estate. He will then place the "Poles of Being," and produce vegetable or animal forms at will.

Thus the prophecy of man's dominion will be fulfilled, for he will have attained knowledge that will enable him to manfacture a psychoplasm (if I may be permitted to coin a word) from which he can bring forth all manner of vegetable or animal life.

> Let man stand upright and splendid,
> Let woman look up from the sod—
> For the days of our bondage are ended,
> And we are at one with God.

Intuition is information direct from the source of all knowledge, vibrating the brain cells and nerve centres of the human organism—the temple, or instrument, of the living God—at different rates, or tones, according to the chemical combination of physical atoms composing the organism. An alligator, a horse, a monkey, or a man being organized each on a different key, receives and expresses the infinite word according to its note, molecular arrangement, or chemical formula.

Wireless telegraphy is demonstrating the underlying principle of what the world has named intuition.

Mental or absent healing is scientifically explained in the explanation of wireless telegraphy. The same substance—air, or ether—fills the so-called space in which we exist. We (our bodies) are strung on this attenuated substance like spools on a string; it extends through us—we are permeated with it as water permeates a sponge.

When the brain cells of the mental healer are acted upon by the word or thought of the healer, they vibrate, jar or oscillate at the rate that causes an arrangement of cells that manifest or materialize the bodily functions on the plane of health. This rate of vibration, started from the sender, will produce the same rate, or jar, in the brain cells of any one attuned to the note, if they recognize the operation, for it is only through such consciousness that we become a receiver.

Thus we realize the truth of the statement, "Thy faith hath made thee whole."

The sender and receiver in the Marconi system must be in the same key—that is, adjusted to sense the same jar, or vibration. When the brain of the healer and patient are in unison, through conscious understanding or agreement, cures can always be effected if the chemical constituents—molecules—are present in the organism of the patient, though dormant, negative, or out of harmonious co-ordination, by the proper jar, or thought vibration, of the healer or sender.

But if the blood of the patient is really deficient in some of the mineral or cell-salts, the phosphates, sulphates, and chlorides of iron, lime, potassium, and other inorganic substances which compose the material organism, the cure cannot take place unless the jar, or vibration, of health started by the sender so oscillates, or jars, the fluids of digestion and assimilation that these lacking elements may be set free from the food and water taken by the receiver (patient), and thus supply the deficiency. It is these cases that baffle mental or divine healing.

Biochemistry fills the gap, offers the solution of the problem by preparing the cell-salts and proceeding directly to the work of doing that which is absolutly necessary to be done, namely, supplying *the necessary chemical molecules.*

The body is a storage battery, and must be supplied with the necessary chemicals, or it will not run. When this can be done by right thinking, well and good; but when a deficiency does occur, why not supply it direct by a biochemic procedure?

A child may touch a button that will start a complex machine to operating, and yet not understand the science of physics or mechanism of the machine.

So there may be many methods of starting the workmen in the system that have become dormant into action. Massage, bathing, electricity, magnetic healing, suggestion, absent treatments, concentration, affirmation, prayer, all these and many more that might be mentioned, can and often do start forces that have become dormant because some link in the chemical chain of inorganic molecules has been misplaced or thrown out of gear, *but when these chemicals (man's body is a chemical formula, remember) are deficient* in the blood, you can no more supply them by any of these modes of operation than you can cure hunger by them. These methods are all good in their time and place to start dormant energies, but none of them will supply deficiencies —viz., cure hunger.

So Biochemistry furnishes the key to all cures made by the old or allopathic, the homeopathic or electric schools, or by medical springs or healing through the operation of mind.

There are some who heal by thought transference, others must come in contact with the patient. In either case, I hold the process is orderly and within the domain of law.

This science is in perfect harmony with the Chemistry of Life operating in each human organism, and cannot antagonize any phase of higher thought. Mind or mental cures, Christian or Divine science, suggestive therapeutics or magnetic healing, must all operate according to Divine law (Life Chemistry), or not at all. The operation of wisdom has many names, but the *chemical process* is one.

END OF THE WORLD ASTROLOGICALLY CONSIDERED

(Written in 1913)

THE word "world" is derived from whirl, to wheel or turn round, to gyrate. In a broader sense, it is used to express activity. Thus we say the financial world, the labor world, the world of music, etc.

About every 2000 years the Solar System passes through one of the Signs of the Zodiac. During a period of 2000 years, ending about the year 1900, the Solar System was in the sign Pisces, the fishes, a water sign. During this time we were in the "flood that covered the whole Earth."

Water causes the appearance called refraction or illusion. Thus a straight stick appears crooked while immersed in water.

"Avoid the *appearance of evil*" has a scientific meaning. The word Jesus is from a Greek word meaning fish.

JESUS, JEHOVAH, JOB and other scientific formulas were never intended for pronunciation as a word, but like the chemical formula of water, H_2O, were intended to express principles.

The word fish in its last analysis, means substance from the sea. Virgin Mary means Pure Sea, or water. In the Pisces age, we had a fish god, quite naturally, and his disciples were fishermen. For 2000 years we were in a *water-world* age. The fish-god walked on the water and fishermen preached his gospel. Then came navigation, Columbus, Watt, Fulton, and finally submarine navigation. For 2000 years water ruled the world. Then the Solar System swung into the air Sign Aquarius. And now it is the airship that rains fire upon water-ships, and "Alas, alas that great city wherein were made rich all that had ships in the *Sea* by reason of her costliness; for in *one hour* is she made desolate." Rev. 18-19.

Aquarius is the only *human* (or man) Sign of the 12 Signs or Sons (Suns) of Jacob. Jacob in Hebrew means circle. The Son (Sun) of Jacob, Dan (Daniel) means a Judge and is the Son that rules Aquarius, therefore we are now in the Judgment age. The "Day of Judgment" will last about 2000 years, or while the Solar System is in Aquarius and ruled by Dan, the Judge.

So then, statements made relative to the end of the world in a general way referred to the entrance of our Solar System into the Sign of Humanity, Aquarius or "Sign of the Son (Sun) of *Man* in the heavens. See Luke 21 and Matt. 24. The direct reference to the real time of the end of the world (whirl or activity) of competition, interest, profit, rent and exploitation of labor, is clearly explained in Rev. 18.

The letters "B. A. B." means "Doorway to God." " Y. L. O. N." means confusion. So Babylon means the Door to God in confusion, or the overthrow of the churches that claimed they taught the way to heaven, and that heaven is somewhere other than where the scriptures teach that it is. "The Kingdom of Heaven is within *you*." Your bodies are the temple of the living God."

"The Holy Ghost dwelleth in *you*."

The "blind leaders of the blind" have fallen into the ditch and the old mind of war and competition is being gathered together as a scroll.

Out of the chemicalizing mass of God's creative compounds, out of wars and rumors of war, out from the strange shapes and ghostly shadows that creep over the face of the earth; out from the trenches of a thousand blood-soaked battlefields where God's loosed thunder shakes the world; where vultures flap and air-ships rain fire; where Armageddon's blood reddens the "horse's bridle" and the war god wallows in the mire of the old world's life-blood, there will come forth from the crystallizing carbon of a material age the diamond of a new faith sparkling with the truths of Science and Philosophy.

The revolutionary planet, Uranus, entered the sign of the "Son of Man" Aquarius, February, 1912, to remain seven years, or until February, 1919. Some astrologers figure that Uranus will not pass from Aquarius until October, 1919.

Here is a memorandum of some of the doings of Uranus. During the French Revolution, this "Son (Sun) of Heaven," (Uranus is from the Greek meaning Heavens), was in the constellation that rules France. In 1776, Uranus was in Gemini, the ruler of the United States. Eighty-four years later, the length of time required for this planet to pass through twelve Signs—seven years in a Sign—brings it back to Gemini and then we had the Civil War.

Uranus overthrew Kings in France, then Kings on this continent, then chattel slavery.

In the year 1912, Uranus entered the Sign of Humanity, a world sign, or Sign of the whole Earth, and astrologers look for world revolution and the overthrow of all Kings, Emperors, Czars, Kaisers and tyrants and war-lords whoever and wherever they be.

Commercialism, wage slavery, child slavery, competition, the liquor traffic and the traffic in the bodies of women have received "their deadly wound" from the righteous wrath of a world-awakened conscience.

To the illumined men and women, to those who have the vision of the New Time reflected on the canvas of their consciousness, the end of the world is the prelude to the Age of Regeneration.

All operations of divine wisdom are three-fold. First, the Heavens declare future manifestations.

Second, these operations appear in the commercial world.

Third, man realizes their ultimatum in his physical organism. And so it has come to pass that as the second act—the world cataclysm, draws to its end that man sees the end of animalism in his own body. Gluttony, drunkenness, sex perversion and "wasting his substance" in riotous living" are all called up to the judgment seat.

Prohibition of alcoholic drinks, food conservation, vegetarian propaganda and clean living comes to the fore everywhere and thus we see the end of the old world of sin and ignorance that has dominated man for ages.

"Behold I make all things new."

Then, when the animal is conquered, "No lion shall be there nor any ravenous beast shall walk there."

THE MARVELS OF THE TWELVE MINERALS OF THE BODY

SCIENCE has demonstrated the intelligence of every atom that constitutes so-called matter.

The intelligence of the inorganic, or cell-salts of the blood is the crowning miracle of the wonders of the human body.

The Phosphate of Potassium (Kali Phos.), is the mineral base of all nerve fluids and therefore a deficiency in this cell salt disturbs the brain centers.

The Phosphate of Lime (Calc. Phos.), in its union with albumen, forms bone. Albumen is the organic substance and lime the mineral, which, by union, manufactures bone tissue. Nonfunctional or imperfect bone is the result of deficient lime salt.

Sodium Chloride (Nat. Mur.), absorbs and distributes water in blood and tissue.

Sodium Sulphate (Nat. Sulph·), eliminates an excess of water: one molecule of Nat. Sulph. having the atomic rate of motion that eliminates, or casts out, two molecules of water. Thus the fluids of liver are kept in normal consistency.

Potassium Sulphate (Kali Sulph.), is the mineral base of oil, and when there is the proper amount of this cell-salt in the blood oil is properly distributed, and health maintained.

The Phosphate of Iron (Ferrum Phos.), molecules are carriers of oxygen through the circulation, form red blood corpuscles, and give strength to tissue that forms the sheath of nerves and walls of blood vessels.

Sulphate of Lime (Calc. Sulph.), builds the cells of the lungs and when this salt is deficient the cells become brittle, disintegrate and are thrown out, producing the condition called tuberculosis.

Phosphate of Magnesia (Mag. Phos.), is a constituent of the white fibers of nerves. A lack of proper balance of this salt causes sharp, shooting pains, cramps or imperfect heart action; these different sensations are words or dispatches from brain centers calling for reinforcements.

Space will not permit a full expose of the marvels of operation of the twelve cell-salts, but a complete story of the chemical action may be found in Dr. Carey's books on Biochemistry.

PARADOXES OF CIVILIZATION

THIS is an age of keen investigation, of truth-finding, and idol-breaking. He who is afraid to investigate for fear some cherished idol will be broken is not a true scientist, and not true to himself.

No length of time ever sanctified anything, and the truth alone sets free. But how shall we explain the seeming contradictions that confront us at every turn?

The inconsistencies and paradoxes in the thoughts and actions of man can be explained only on the hypothesis that one power, principle or cause does all and is all, and that so-called paradoxes are but steps in the operation of wisdom, moving in orderly procedure to the completion of certain phases of expression.

In chemistry, acids and alkalis are paradoxical; being opposite, they cause chemicalization, or fomentation, which seems to be confusion. In ancient alchemical writings, this phrase when applied to human life was called Babylon, which, when traced to its root, we find means confusion.

The world's civilization, since recorded history began, has been one chain of paradoxes. The scales of competition forever tipping cause the phenomena. The producers of food starve because they cannot buy the food they produce. The makers of clothing are in rags because they can't pay, in the coin of the realm, for the clothing they make. Prices are too high for the consumer, or too low to please the producer. If workers demand the full product of their labor, they are denounced as thugs and anarchists by those who do not produce anything. In competitive thought there appears to be many inconsistencies.

Man lays his sceptre on the stars, analyzes their substance, and then dies from the effect of acid in his blood, because he does not know what to eat. He foretells the return of a comet to an hour, but cannot tell if he himself will have la grippe next week. He can tell you the hour in the day one hundred years hence that there will be high tide at Bombay, or along the coast of Norway, but he doesn't know the cause of smallpox, and foolishly thinks the decaying organic matter or pus from a sick calf injected into the blood may somehow prevent it.

He can clothe himself in armor and dive to the ocean's floor, or travel three thousand leagues under the sea in a submarine boat, and then be killed by a street car or automobile in broad daylight on the level road. He knows how to keep the chemicals properly balanced in the storage battery of his automobile, but puts alcohol, morphine and tobacco in his own body and wrecks it.

He can tell all about the moons of Jupiter, the rings of Saturn, the transit of Venus, the canals of Mars, and can talk with the man in the moon, but he knows no more about the real composition of his own blood or nerve fluid or the mysteries of digestion or assimilation or the chemical formation of bile, than a politician knows of the true science of government.

Man condemns cruelty to animals, but the slaughter house disgraces civilization, and man expects beefsteak for breakfast. He preaches humanitarianism, but the sweatshops still remain a bloody blotch on the face of humanity.

Patriots, real statesmen, humanitarians, seers, prophets, the far-seeing, self-sacrificing lovers of humanity, are derided, abused, persecuted, imprisoned, tortured, crucified, and then monuments are erected with cold chiseled marble to mark their burial place; and the poet and historian vie in singing their praises, while their features are preserved on innumerable canvasses, and their sightless eyes stare out from bronze and plaster of paris in every library, or stand sullen and silent in a niche in the Hall of Fame.

Men and women teach the everlasting truth that we are all children of one common mother and father—the eternal Positive-Negative Energy from which all things proceed—and therefore members of one family—brothers and sisters in truth—but they insist upon an introduction before speaking to each other as they pass. Sons of God, and yet they must not greet each other without an introduction!

O the pity of it! O the shame! Man asserts that God is love, and then writes in a book, "the fear of God is the beginning of wisdom."

Love is altogether lovely and no one can fear it. Some people are so afraid that they will go to hell when they die that they live in hell all the time on earth. Some are so afraid of smallpox that they poison themselves with vaccine pus, which is more deadly than smallpox.

A woman will express pity and sorrow for the poor and needy, and then put a ten dollar collar on a fat, pudgy dog, and let it lead her along the street, while five orphan children are ragged and hungry in her block. Women love birds—especially on toast; they love the beautiful plumage of birds—especially on their hats.

Man declares the laws enacted by legislatures and congress are sacred. He then violates these laws, carries his case to the court of last resort, and gets the sacred laws repealed. Man declares majority should rule, but bitterly opposes the majority when contrary to his opinion.

Many persons speak of "the ignorant foreigner"—a man who probably speaks three languages correctly, while his critics can't speak their own language as well as a parrot.

The automobile driver may scorch his wagon along the street at a forty-mile clip, if he will toot a horn to warn pedestrians to flee from the wrath coming, but the man on foot is locked up in jail if he runs amuck in the street, shouting, "Get out of my way!" Men say that God is everywhere, and that the devil is everywhere, too, and then prove mathematically that two objects cannot occupy the same space at the same time.

Spiritualists say that the orthodox belief in a personal devil, one big fellow with hoofs and horns, is quite absurd and amusing, and then these reformers declare that most sickness is caused by evil spirits.

A highwayman murders a man for his money, but refuses to eat the dead man's lunch of meat because he must keep the Lenten days.

Physicians experiment with poison, and find that a certain drug will cause disease in the human body, when taken internally; and when called to prescribe for the sick, they administer the drug that causes disease, believing that two diseases are better than one.

The average man declares that woman is far superior to man in perception, intuition, and judgment in regard to social life and the welfare of the family, but he bitterly opposes the franchise for women. Man is not willing that mothers should have a vote in making laws to govern their own sons and daughters. Man seems willing to give to women everything under the sun that they don't need, from chewing gum to china vases, but refuses to give them the thing they need most—the ballot.

Our wives, mothers and sisters may perform all sorts of labor, study all the arts and sciences, own property and pay taxes, and be amenable to man-made laws; yea, a woman may go down into the dark valley, where God's creative compounds materialize in human form, and kiss the white lips of Pain before she holds a babe to her breast, but she must not have a voice in laws to govern that child.

The cry of "hard times" is always heard. People dole out nickels or pennies grudgingly for charitable purposes, or civic improvements, or good roads; but a prize fighter receives a hundred thousand dollars for hitting a fellow on the jaw, men pay five dollars for a bottle of champagne, and two dollars for a porterhouse steak; and a woman will spend forty dollars for a hat that has no more correspondence to the contour of the human body than politics has to honest government, or the co-operative commonwealth.

Men complain of unjust tariffs and taxes, call upon Heaven to witness the iniquity and dishonesty, graft, thievery and fraud of courts and law-makers, and then go meekly to the polls and vote for the same fellows over again.

Tired men and women will squeeze into crowded cars, and step on each other's toes while holding straps; but if you mention municipal or national ownership of railroads as a remedy, they say: "O, I do not bother about politics. Which team won the football game?"

Dogs are allowed to roam at will over lawns and amongst the flowers, while little boys and girls stand on the walk and read a sign board, with this strange device, "Keep off the grass!"

Man must be quiet and orderly, must not talk, or laugh in a manner that will disturb the peace of his neighbor; but dogs may bay at the moon and planets, the comet, the dog-star, the Pleaides, or any old thing; and cats may curse and swear, and rip the boards off the woodshed in an unearthly noise contest, and the poor wretch who dares to protest is called "a cruel, bad man!"

A citizen of the United States reads about the Alps, buys an alpenstock and a ticket for Europe, spends three years abroad, and incidentally spends sixteen thousand dollars. But when the Swiss peasant asks the United States citizen about Yosemite, or Shoshone, or Niagara, he is oftentimes amazed to learn that the American tourist has never worshiped at the feet of El Capitan, that he never thrilled with awe as he looked at the descent of Snake River into the awful gorge at Shoshone, nor ever bathed in the rainbow mists above tumbling Niagara.

Think of a man going to Europe to see the sights, who has never looked down into the riven earth, where the Colorado canyon reveals Nature's carvings and colors, nor passed through the Enchanted Gateway of the Cascade Mountains, where Jupiter Olympus hurled thunderbolts in the anicent days, and dug a canal between the snow-capped mountains, Adams and Hood, and let the inland lakes flow to the Sunset Sea!

"See Columbia's scenes, then roam no more; naught else remains on earth to cultured eyes." Columbia, the "river of the west." The Nile might come from its cloudy heights and pour the water of Egypt into this mighty stream, and it would cause no ripple upon its broad expanse, nor would it increase the speed in its stately march to the sea.

The firs of Oregon and Washington and the redwoods of California equal the Cedars of Lebanon; and the pillars of salt on the shores of Palestine's Dead Sea is outdone by the dead sea on Utah's plains. Truly, the man is a paradox who explores every country on the globe, except his own.

The Christian professes to believe that after the change called death of the body, his soul will be wafted to a place called Heaven—a place of beauty and eternal peace, where he will have a mansion of gold, windows of precious stones, fronting the pure waters of the River of Life, that flows from the Throne of God; but let that same Christian have a pain in his stomach, and he sends a hurry call to the nearest physician and begs him to use all his skill to keep him here in this vale of tears—here among the "beggarly elements of the world," where he may remain a "worm of the dust" a little while longer.

Mental Scientists affirm health and opulence, and then fail in business and change climate to cure nervous prostration.

Of all the negative conditions—devils—the race is subject to, Fear is the greatest. We are born cowards. Our mothers feared for us before we were born. We came into earth-life with a wail of fear. All who had anything to do with us feared something evil would happen to us. They were afraid we would catch cold, or the measles, or whooping cough, or diphtheria, or die of "summer complaint." Somebody feared all the time that we would get scalded or frozen, or fall out of bed, or down stairs, or into the well.

When we were old enough to be afraid, we feared our parents, our teachers, the ministers, the dark, the devil, and even feared God, whom St. John says is Love. Later, we were afraid in business, of fire; afraid the election would start some one to tinkering with the tariff, or our blessed money system. We were afraid on land and sea, of fire and water, cold and heat, wind and hail, lightning and cyclone, earthquake and tidal wave, and yet we wonder why there are so many sick people. But the silliest of all fears is the fear of microbes.

We laugh at the elephant because it fears a mouse, but the ignorance of the elephant in that respect is pure wisdom when compared with man's fear of contagious diseases, and his senseless efforts to "stamp them out," by quarantine, disinfectants, germicides, lymphs, serums, and vaccine pus galore.

Paradoxes upon paradoxes! Yet all will cease when competition is supplanted by co-operation. So, then, to sum up, we must find the reason for competition. Man has fostered competition because he thought he was an individual. Man has turned the mighty power he possesses to every object and principle or force in the universe, except himself, the greatest miracle of all. When man focuses his divine thinking lens upon himself, he will realize that he is an epitome of unlimited Cosmic Energy. Then the "Heavens will roll together as a scroll" and reveal the Real Man as "the Lamb of God that taketh away the sins of the world."

I have seen the surface of flowing rivers changed to ice by the chemical action of cold. I have seen this crystalized water break into countless pieces by the action of heat. I have seen this grotesque, jagged army rush down the great waterways like a charge of cavalry and sweep away iron-girded bridges that an hour before had safely borne the traffic of commercialism. I have seen these huge blocks of ice beat and batter and heard their grinding crash, like the hammer of Thor at the gates of the imperial cities of civilization. I have seen the surface of human thought smooth and petrified by inaction, conservatism and respectability. Under the influence of love and wisdom, or in the presence of a great necessity, I have seen this petrification break into projectiles of dynamic thought, dominated with some new concept of life, and hurl themselves against the mailed forces of error and prejudice, until domes and citadels and towers reeled and fell.

As the blocks of ice finally melt and become one, with their primal elements, so will diversity of thought finally unite and blend by the Chemistry of Energy into one harmonious whole.

"THE NEW NAME"

"And I will write upon him the name of my God."
"And I will write upon him my NEW NAME."—Revelation.

Man struggling up to the sunlight,
  Up from the mire and clay,
Fighting through wars and jungles,
  And sometimes learning to pray—
And sometimes a king with a scepter,
  And sometimes a slave with a hod—
Some people call it Karma,
  And others call it God.

A beggar ragged and hungry,
  A prince in purple and gold,
A palace gilded and garnished,
  A cottage humble and old—
One's hopes are blighted in blooming,
  One gathers the ripened pod—
Some call it Fate or Destiny,
  And others call it God.

Glimmering waters and breakers,
  Far on the horizon's rim,
White sails and sea gulls glinting
  Away till the sight grows dim,
And shells spirit-painted with glory,
  Where seaweeds beckon and nod—
Some people call it Ocean,
  And others call it God.

Cathedrals and domes uplifting,
Spires pointing up to the sun,
Images, altars and arches,
Where kneeling and penance are done—
From organs grand anthems are swelling,
Where the true and faithful plod—
Some call it Superstition,
While others call it God.

Visions of beauty and splendor,
Forms of a long-lost race.
Sounds and faces and voices,
From the fourth dimension of space—
And on through the universe boundless,
Our thoughts go lightning shod—
Some call it Imagination,
And others call it God.

Acids and alkalis acting,
Proceeding and acting again,
Operating, transmuting, fomenting,
In throes and spasms of pain—
Uniting, reacting, creating,
Like souls "passing under the rod"—
Some people call it Chemistry,
And others call it God.

Vibration of Etheric Substance,
Causing light through regions of Space.
A girdle of Something, enfolding
And binding together the race—
And words without wires transmitted,
"Ariel"-winged, spirit-sandaled and shod—
Some call it Electricity,
And others call it God.

The touch of angel fingers
  On strings of the human soul.
Anguish and ripples of laughter
  Written across its scroll.
Chords from the holy of holies—
  From sunrise sky to the sod—
Some people call it music,
  And others call it God.

Earth redeemed and made glorious,
  Lighted by Heaven within;
Men and angels face to face,
  With never a thought of sin—
Lion and lamb together
  In flowers that sweeten the sod—
Some of us call it Brotherhood,
  And others call it God.

And now the sixth sense is opened,
  The seventh embraces the whole,
And, clothed with the Oneness of Love,
  We reach the long-sought goal—
And in all Life's phases and changes,
  And along all the paths to be trod,
We recognize only one power—
  One present, Omnipotent God.

PART TWO

"For he that soweth to his flesh shall of the flesh reap corruption; but he that soweth to the Spirit shall of the Spirit reap life everlasting."

—*Galatians 6:8.*

PHYSICAL REGENERATION

THE inner eye—"the eye behind the eye"—just above and attached to the pineal gland by delicate electric wires, or nerves, is called Optic Thalmus, and means "Light, or Eye of the chamber."

In the Greek, it means "The light of the World." "The Candlestick," "Wise Virgins," "The Temple Needs no light of the Sun," "If thine eye be Single, Thy Whole Body shall be Full of Light," and other texts in the New Testament refer to the single eye or Optic.

Let us now search for the oil that feeds this wonderful lamp, the All Seeing Eye.

Christ Jesus is made to say "I Am the Light of the World." The word "world" comes from "whirl," to turn as a wheel, to gyrate, etc.

The human body is a certain rate of activity, motion or whirl, *i.e.*, world, and light of the world and the temple that needs no light of sun or moon refer to the body "Temple of God," when there is "oil in the lamp."

Error is not sanctified by age. It behooves every lover of truth to cast aside prejudice and dogma and find truth.

Until we know the meaning of the words "Jesus" and "Christ" we will not understand the bible which was written in Greek and Hebrew and translated and retranslated to suit the whims and ignorance of priests and charlatans all down through the centuries.

Constantine, a beast in human form, who murdered his mother and boiled his wife in oil, was the chief factor in the orthodox translation of the so-called King James bible.

Constantine was told by the priests of his time that there was no forgiveness of crimes like those that he was guilty of and so this Roman Emperor devised the plan of salvation in order that the blood of the innocent Jesus (or Christ) might save him from eternal damnation. An easy way out for this monster, and all the other blood-smeared tyrants, Kings, Emperors and Napoleons of finance, competition and war, from Pharaoh to the present-day rulers whose thrones and scepters lie scattered and broken along the Highway of Nations (1917).

"Here the vassal and the King, side by side, lie withering. Here the sword and scepter rust. Earth to earth and dust to dust." The word Jesus is from Ichthos, Greek for fish. The word "Christ" means a substance of oil consistency, an ointment or smear. Varnish or paints are used to preserve or *save* wood or paper or cloth—hence they become Saviors.

At about the age of twelve, Jesus was found in the temple arguing with the doctors or teachers. The word "doctor" is from Latin "docere," to teach.

Every month in the life of every man or woman, when the moon is in the sign that the sun was in at the birth of the individual, there is a psycho-physical seed or "Son of Man" born in the Solar Plexus or the pneumo-gastric plexus which in the ancient text was called the "House of Bread."

Bethlehem, from Beth, a house, and helm, bread. "Cast thy bread upon the waters and it shall return to thee after many days." Waters are the blood and nerve fluids of the body that carries the fish on its "Divine Journey" to regenerate, save and redeem man. Nazareth means to cook. Nazarene means cooked. Cook means to prepare. Any materialized thing is bread, Nazareth, mass, maso, or dough. Thus the Catholic Mass. Also Maso-n. It will now be made plain why the Masons and Catholics are not in agreement, for our letter N is an abreviation of the 14th letter of Hebrew Alphabet, Nun, a fish. By adding N to Maso, the riddle of cooked or prepared fish was made so plain that the priesthood strenuously objected, and thus developed friction between the church and Masonry.

The disciples were fishermen. The early Christians used a fish as their secret symbol. Money to pay taxes was taken from the mouth of a fish. Bread and fish were increased until twelve baskets full were left, etc. God prepared a fish to swallow Jonah. Jonah means dove. Dove means peace—the germ descending from the gray matter of the brain (See baptism of John). The storm means sex desire. The life seed was thus saved. "He that is born of God cannot sin (or fall short of knowledge) for his seed (fish) remaineth in him."—John. The age of puberty is about twelve. Up to that age, a child does not understand moral responsibility. "The first born" means the first seed or fish. Pharaoh, sex desire always tries to destroy the first born. When Jesus was born in Bethlehem, he went (at the age of twelve) up the pneumo-gastric nerve which crosses the medulla oblongata at its junction with spinal cord at the head of the "River of Jordan," the marrow or nerve fluid of the spinal cord (See illustration in Physiological Charts) and enters the cerebellum, the temple. This is the temple where the moral seed argued with the purely animal cells to change their rate of vibration to moral and spiritual concept. Later this seed (Jesus) drove those who bought and sold ("Even as you and I") with a whip of thongs out of the temple. We must all give up the animal life or suffer the same fate.

Before we explain the baptism in Jordan and the christening and the crucifixion, etc., let us briefly explain Moses, Joshua, Nile, Pharaoh and the children of Israel.

Egypt means the dark lower part of the body. That part of the body below the Solar Plexus is Egypt, or the Kingdom of Earth. All above the center constitutes the Kingdom of Heaven. ("The Kingdom of Heaven is within you.") The Manger, or Bethlehem, is the center, or the balance.

Nile, Moses and Pharaoh's daughter, all refer to generation. (See overflow of Nile). It rises in the mountains of the moon. Moses means "drawn from the water." Fish are drawn from water. "There are *two* fishes in our sea"—Vaughn. See Sign of Pisces, two fishes.

"Joshua the Son of Nun." Nun is Hebrew for fish.

Moses was the physical or *generative* fish.

Moses' laws were on the *physical* plane.

Joshua means "God of Salvation," and salvation comes from saliva or *salivation*. Sal is salt which *Saves*. "If the salt loses its Savor" *i.e.* Savior, wherewith shall it be salted?" Saliva *saves* the body by digesting (or preparing) the food. Saliva is a smear or ointment, and so Joshua compares with Christ as Moses compares with Jesus. Moses died on Mt. Nebo. Nebo means understanding. Joshua took the place left vacant by the death of Moses. Jesus was baptized *of* John in Jordan—the fluids, Christ—substance of the spinal cord and became "my beloved Son in whom I am well pleased." There is no J in the Greek or Hebrew alphabets, therefore, the word "John" I O H N means "Soul" or "fluids of the body" and not the Ego or Spiritual Man. So when the body dies, the fluids die—thus man loses his soul when he loses his body. To prevent the loss of soul and flesh is the mission of the Son, or Seed, of God, or the Son of man. There are two very small nerves that extend up from the Sacral Plexus on each side of the spinal cord and cross at the base of the brain and unite the Medulla oblongata. These nerves or delicate wires are called Adi and Pingali. After the seed or fish has been *Christed,* if it is retained and not wasted in sexual desire, it goes up to "Galgotha" the place of the skull" and crosses the wires, then remains three days in the tomb or the three chambers of the Pineal gland, then it enters the optic, or third eye, and "giveth light to all that are in the house," that is the beth or body, all the twelve functions represented by the twelve disciples—the twelve signs of the Zodiac.

But the question will be asked—What or where is the source or origin of this seed or redeeming Son? We answer: "Ether, Spirit or God." Names mean nothing.

Esse, Universal intelligence, or *It* may be used. *It* breathes into man the breath of life. This elixir is carried through lungs into arteries, or air carriers, where it unites with the inorganic cell-salts, materializes (cooked) and forms granules, and is then deposited as flesh and bone.

The study of Astrology, Biology and Biochemistry, added to Physiology, will lead one into the great Alchemical laboratory of the "Fearfully and wonderfully made" human temple—the temple made without sound of saw or hammer.

Before the Neophyte can fully realize the power of the Divine Eye within his own brain, he must understand the meaning of *Or* especially in its relation to *Word* and *Jordan.*

Iordan (not Jordan) is the word in the original text. I is from Iod—the 10th letter of Hebrew Alphabet, and means "hand" or that which creates. *Or* is gold, not metal, but the "precious substance"—the seed. Dan is Hebrew for Judge, therefore the Creative Power operating through the precious substance produces Judgment, or the man of good judgment.

The upper brain is the reservoir of this *Or* and is the gray matter or "Precious Ointment" or Christ.

"In the beginning was the *Word* and the *Word was* God. *All things* were (or is) created by it" etc., etc.

The "Lost Word" is a symbol of the generative or animal thought eating the fruit of the Tree of Life—thus destroying or losing the gold, "or," of the body.

Hiram means "high born," or the seed destined to reach the pineal gland and "Single Eye."

Tyre means a rock. By the conservation and transmutation of the sex substance the pineal gland becomes firm and hard and is, in the fable, called rock (Tyre). "The wise man built his house, Beth or body, upon a rock." So here we have the explanation of Hiram Abiff. Abiff is derived from the word Abid—month.

Hiram Abiff (there are some who will understand) was resurrected during the delay caused by searching for his body: in other words 28½ days passed (a *month*) and another seed was born which the candidate for initiation is admonished *not to slay.*

The upper brain is the *Word* and it furnishes *all* that man contains, or is. Jesus was not a Savior until he was *Christed of Iohn* in the *Jordan.* Then he became the "Beloved Son."

Why was the baptism necessary? Because there are two fish, one was Jesus the Carpenter, the man. The other, the *Christed* Jesus, the Son of God. The Christ substance gave the electric or magnetic power to the seed to cross the nerves at Galgotha without disintegrating or dying.

To crucify, means to add to or increase a thousand fold. When electric wires are crossed, they set on fire all inflammable substances near them. When the Christed seed crossed the nerve at Galgotha, the vail of the temple was rent and there was an earthquake, and the dead came forth, *i.e.* the generative cells of the body were quickened or regenerated.

The crucifixion or crossing of the life-seed gives power to vibrate the pineal gland at a rate that causes the "light of the chamber" to fill the "whole body with light" and send its vibration out along the optic nerve to the physical eye and thus heal the blind.

Let us hark back to the Nile: Pharaoh means a ruler or a tyrant or sex desire. Israel means blood, children of Israel, molecules of the blood, but more particularly refers to the thirteen monthly seeds born in 365 days.

THIRTEEN, THE OPERATION OF WISDOM,

THE number thirteen is unlucky for ignorance only.

All so-called laws of nature may be reduced to thirteen.

The origin of words and their application vary widely. Thus the origin of twelve is circle or completeness, or without break or sin; that is complete. All operations that produce something may be called twelve, being complete in order to produce, the product is therefore thirteen. Thus all machines or factories symbol twelve and the product thirteen.

The Zodiac

There are twelve constellations, the central suns of which constitute the signs of the Zodiac.

One sign rises every two hours, or so appears to our sense, because the rotation of earth causes the phenomenon, and the earth, or sun, makes thirteen.

The Human Body

There are twelve functions of the human body and the seed, or psycho-physical germ, born in the solar plexus every 28½ days makes thirteen. So then there are thirteen moons in 365 days. The pneumo gastric nerve, vagus nerve, that comes down from cerebellum across (a cross) the medulla oblongata branches out at the lungs (pneumo) and at the stomach (gastric), and is called "The Tree of Life" (thirteen letters), also pneumo gastric (thirteen letters).

There are twelve mineral salts in the blood and blood itself—the product—thirteen.

David's Thirteen Sons

1st Chr., 14th ch., 3d ver. "And David took more wives and concubines at Jerusalem; and David begat more sons and daughters."

Here follows the names of thirteen children.

Thirteen Children of Jacob

The 29th and 30th chapters of Genesis record the birth of eleven sons and one daughter, Leah. The 35th chapter records the birth of Benjamin, the 12th son and 13th child.

Jacob, in Hebrew, is circle, or to follow after, also represented in Hebrew symbology by a circle of men, each one with hand holding the heel of the one in front, and thus describing a circle.

The origin of the allegory is founded in the rotation of earth and the apparent rising of one of the Signs of the Zodiac every two hours, making twelve, and the earth itself thirteen.

The esoteric meaning is based in the marvelous operation of the wonderfully made human body. All of the parables, fables or allegories of the human organism are related to 13.

Moses, Joshua, Jesus, Christ and all the characters of the scriptures are symbols of the psycho-physical seed that is born in, or *out* of the solar "manger" in the center of the body, fed by the delicate nerves that branch from the "*Straight* and narrow way"—the pneumo-gastric nerve (or vagus nerve), so-called because there are so many wandering branch nerves called the branches of the "Tree of Life."

Out of this receptacle or manger a seed is born every 28½ days, or *thirteen* every 365 days. The birth ("first born"—or born first), occurs in the life of each male or female, when the moon enters the sign, each month, that the sun was in at the birth of the individual.

Twelve symbols a circle, in Hebrew, meaning complete. The product of twelve is thirteen. Gallilee is a circle. The Sea of Gallilee, circle of water, or fluid, hence circulation of the blood and fluids of the body. So Jacob may be applied to the body.

Rachael means Ewe, or Mary, Eve or the manger (solar plexus) where Mary and Jesus were found. There is no U in the Hebrew alphabet, hence no double U. So V is the letter, or double V—hence EVE or EVVE—*i.e.* Rachael. The solar plexus is symboled by many names in the Bible, all female, whether they refer to a man or a woman, because it gives birth to the seed. This wondrous redeeming seed is exactly the same in male and female and plays no part in generation, but is the "Plan of Salvation" whereby the child "born in sin" may be redeemed and saved. Thus, "In my Kingdom (regeneration) there is no marrying," etc.

For key to Benjamin, the 13th, the entire chapter, Genesis 35th, should be studied carefully in the light of the new revelations.

Sixth verse, 35th chapter: "So Jacob came to Luz, the same is Bethel" (or Beth-lehem), house of bread, the solar plexus. "He built there an altar (same as "manger" or plexus,—womb) and called the place "El-beth-el," (God's house of God) because there God was revealed, etc.

Sixteenth verse: "And they journeyed from Bethel and there was some distance to come to Eprath (fruit, posterity, Bethlehem, seed).

Here Rachael "had hard labor" and gave birth to Ben-oni and died. Ben-oni means "child of my sorrow," but Jacob called him Ben-Ja-min, "Son of my right hand."

Sixteenth to twentieth verse: "And Rachael died and was buried in the way to Ephrath (the same *is* Bethlehem). "And Jacob set up a pillar upon her grave; "The same is the pillar of Rachael's grave unto this day." The solar plexus, chamber or manger is the pillar.

The death of Rachel, the mother, simply means that thirteen completes the number of seeds born during the thirteen moon months.

Great latitude must be given to writers of parables, fables and allegories.

Genesis, 35th chapter, 10th verse: "And God said unto him, thy name is Jacob (the circle), thy name shall not be called any more Jacob, but Israel shall be thy name."

Israel here clearly points to the seeds, thirteen every twelve months, that cross Jordan. Twelfth verse, 35th chapter: "And to thy seed after thee will I give the land."

Jacob (circle) means complete operation; and thirteen the seed, Israel, the product.

Joshua and Jericho

In the book of Joshua (Son of Nun), a fish—born in the solar plexus thirteen times in 365 days, it is recorded that the host marched around the walls of Jericho once daily for six days and seven times on the seventh—thirteen.

Jericho, lately captured by the British troops, 1917, is situated thirteen miles outside the walls of Jerusalem.

Jesus and the Twelve Disciples

Before the crucifixion of Jesus, the seed, fish, there were twelve Disciples or workers and Jesus was the thirteenth. After the crucifixion, which means to *increase in power,* (note the increase in power of the electric current when the wires are crossed), Paul was added to the twelve Apostles. Paul is made to say: "I was born out of time."

The meaning of Paul is small or the "still small voice," as P is from the Hebrew letter Pe, to speak, or the mouth. S is from the 21st letter of the Hebrew alphabet, Schin, meaning falling short of completeness as there are 22 letters in the Hebrew alphabet. So the allegory makes the allegorical character Saul before conversion, or regeneration, and Paul, the *preacher,* after the transmutation.

Smith's Bible Dictionary admits that there are no dates to the so-called Epistles of Paul. Neither are there dates to *any* of the writings—scriptures—gathered by the Council of Nicæ under the Pagan Emperor, Constantine. No one knows when they were written.

The United States and Thirteen

The thirteenth degree of the Zodiacal Sign Cancer was rising July 4th, 1776, when the Declaration of Independence was signed. Cancer represents the breast and is therefore the mother sign, or woman. M is from Mem, the thirteen letter of the Hebrew alphabet, and means *woman.* The United States plays the part of mother to all peoples and gathers them under her protecting care. We commenced our individuality as a nation with thirteen states.

In 1782 the obverse side of the United States Seal was made and contained thirteen stars, thirteen stripes, and an eagle with a quiver containing thirteen arrowheads in one tallon and an olive branch with thirteen leaves in the other. And the motto "E pluribus unum" contains thirteen letters.

About this time, 1782, an unknown man appeared in Philadelphia and offered the drawing of a seal (see cut) which he suggested be added as the reverse side. This man declared that the seal would be adopted in the Year 1921, the digits of which equal thirteen, and that the eagle would no more be used.

Strange to say the stranger's seal was adopted, but has not come into prominence until within the past three or four years.

The reverse side of the United States Seal, as shown in the cut, shows part of the pyramid of Egypt, the base of which covers thirteen acres.

There are thirteen steps or terraces. The motto over the pyramid, "Annuit Cœptis," contains thirteen letters and is Latin for "Prosper us in our undertaking."

Our solar system has passed out of the water sign Pisces, and thus occurred "the end of the world"—thirteen letters.

So our great fleet of planets and flag ship Sun is now in the air or spiritual sign Aquarius.

In the allegory of the suns or sons of Jacob (see Genesis 29), the fifth son born was Dan, a judge, thus Daniel—"Judge appointed by God," as El is face of God in Hebrew.

The first son or sun was Reuben or Libra, the loins, therefore the fifth would be the legs, or Aquarius, sign of man, where the solar system is now and where it will remain for over 2000 years.

Day means an indefinite period of time—thus we say Napoleon's day or Lincoln's day. Therefore, it is plain to be seen that we are now in the "Day of Judgment," thirteen letters.

And once more: Now amid the smoke and fire, the roar of guns and stabs of the submarine; now while thrones crumble and towers fall; now while Armageddon's battle blood fills a thousand miles of trenches and the thrones and scepters lie scattered and crushed along the highway of nations, there sits in the presidential chair of America, whose emblem of flaming stripes and blazing stars is leading the world up to the sun-crowned peaks of Brotherhood, the most remarkable man of all time,

WOODROW WILSON

and the number of letters in the immortal name is

THIRTEEN

And so it shall come to pass, on the morrow, that America will sit at the "Peace Table" and then head the procession of nations redeemed and march on to the Promised Land of the New Order sandaled with righteousness and crowned with victory and rejoicing. So be it.

(The above was written in March, 1918.)

DANIEL IN THE LIONS' DEN

THE word Dan, in Hebrew, means a judge. Dan *iel,* judgment or God's judge. El or iel, in Hebrew, represents the supreme ruler or God. God and good are synonimous, *i.e.* Daniel—good judgment or wisdom.

The word Darius, traced to its root, simply means an office, same as Presidency, and whoever fills the office is for the time called Darius.

Medes is from media, the middle, and is represented in the body by solar plexus.

Persia, the East, Persians, people of the East.

In scripture allegories East always means the back; West, the front; North—up, or the head; South, down—the feet.

There is a wide difference between the original meaning of a word and the multiple applications of a word. For instance, lamb, dove, hog, wolf, eagle and names of all birds and animals represent ideas or principles that have been applied to different species of animal forms on the hypothesis that these names fitted some peculiar trait or habit of the animal or reptile to which they gave the appelation.

Lion means strength and is used to designate the "King of beasts" or animals. The part of the human body below the solar plexus is referred to in the Scriptures (physiological writings) as Kingdom of Earth, hades, lower Egypt and the seat of sex desire (Pharaoh) or the animal passions, appetites, etc.

Ani: breath or soul.

Mal: bad, or imperfect, hence malformation, malnutrition, bad breath or soul (unregenerated substance), or the Ani*mal* man.

The "Lion's Den" is used in the fable to typify the animal functions that were regenerated by wisdom or good judgment—Daniel.

The following definitions will assist the reader to more fully realize the esoteric meaning of words in scripture:

DANIEL—Judgment.

BELSHAZZAR—Bel, Belial or Be-elzebub, has formed a king.

BELTESHAZZAR—A maintainer or Prince . (This title was given to Daniel after his regeneration as shown by the letter T from Tav, the 22nd letter of the Hebrew alphabet meaning cross, where the redeemer (seed) is crucified. "There is no name under heaven whereby ye may be saved except Jesus, the seed, *Christ*-ed and crucified."

NEBUCHADNEZZAR—From Nebo—understanding. A protector against misfortune.

ELAM—Unlimited duration.

A-BEDNEGO—Servant of Nego, *i.e.* understanding.

MESHACH—Guests of Sha, the Son-god.

SHADRACH—Royal or rejoicing in the way.

(Thus it is made clear why Shadrach, Meshach and A-bednego were not consumed in the fiery furnace. They are principles, eternal verities that are not affected by physical expressions and can, therefore, complete the initiation of the Ego).

In the 8th chapter of Daniel, verses 1 and 2, we find the words "Shushan the palace which is in the province (or country) of Elam; and I was by the river Ulai."

ELAM—Eternity.

SHUSHAN—From Susanna, a lily (known as the Capitol of Elam), real meaning, the product of divine mind.

ULAI—From Hebrew Pehlvi, meaning pure water.

Daniel was "By the river Ulai." Ulai here refers to the spinal cord. The marrow, or oil, in this channel is pure crystal in color.

"And he showed me a river of water of life, bright as crystal, proceeding out of the throne of God and the Lamb, in the midst of the street thereof. And on this side of the river and on that was the Tree of Life bearing twelve manner of fruits yielding its fruit every month; and the leaves of the tree were for the healing of the nations."

Month is from Moon (Moonth) and there are thirteen moons in the Solar year.

Leaves are effects of a tree.

The monthly seed (fruit) when saved, not "eaten," heals disease and sin.

"Moses lifted up the serpent in the wilderness," the *body.*

Moses, the first born, the seed, desired to regenerate the blood and lead it to the promised land, thus he lifted up the animal forces, sex desire, here symboled as the serpent (see the temptation of Adam and Eve) . . . So shall the "Son, or Seed, of Man be lifted up," etc., etc.—that is put on the cross in order to reach the pineal gland.

"If I be lifted up, I will draw all *men* unto me." I will draw all other seed unto me. Study the etymology of "men." Also read "The tree in the midst of the garden bore fruit every month and its leaves were healing."

The Commandment to not *eat* of the *fruit* of this tree was not (is not) heeded by the race and death is the result.

The serpent said "Eat, thou shalt not die," but sex desire was a liar from the beginning.

A noted Professor of Greek in one of our universities says that the translation of many New Testament texts from Greek are radically wrong. For instance, "He that saveth his life shall lose it, and he that loseth his life, for my sake, shall find it," should read: "He that saveth his seed—life—shall *loosen* it (set it free), and he that *loosens* it, shall find it," which means that this "Bread cast upon the waters" shall redeem him. Gallilee means a circle of water—the fluids of the body.

Jesus walking on the water is a symbol of the seed, or fish, on its journey. Peter, from petra (stone) is a symbol of physical or material thought which was rescued by the fish, Savior.

The Optic Thalmus, or light in the room, is called "The

Lamb of God that taketh away the sins of the *world.*" The Hebrew letters Lamed, Aleph, Mem and Beth form the word Lamb, meaning innocence or purity.

Sin is from the Hebrew letter Schin, meaning to fall short of knowledge. Sin does not mean wrong or crime, but one may commit a crime or do wrong through lack of knowledge. Paul said: "I die daily" . . . I am the chief of sinners." Revelations: "And the lamp thereof is the Lamb." The word "Lamb" ends with B, which means a house or body of some kind. Now, the optic or central single eye is a body, like the outer eye ball, therefore, a beth. This is called lamb by the ancient poet.

Lamp ending with P, which means speech or *sending forth* or radiating, is from Pe, the 17th letter of the Hebrew alphabet, and was used to express light or knowledge emanating or going forth from this eye or "Lamb of God."

"As a man thinketh in his heart, so is he."

The cerebellum is heart-shaped, and in the Greek is known as the heart. The organ that divides blood was called the "Dividing Pump." The seat of thought is the Cerebellum. Our thoughts shape our lives. If we think continually below the solar plexus in the Kingdom of Earth; if we dwell in thoughts of material pleasures, we become animal and materialistic. If we really desire the Kingdom of Heaven, we must think of the process that will enable us to realize it.

When Jesus was born, they put him in "swaddling clothes." Now the psychic germ (fish) is composed of the concentrated essence of life and is covered by a gossamer capsule for protection. If this swaddling cloth is broken, the "precious ointment" is lost, *i.e.* it disintegrates and corrupts the blood.

In order to save this germ of life, man must remember that as a man thinketh, so is he. While men must abstain entirely from sexual contact, he must also realize that "He who looketh on a woman to lust after her, hath committed adultery with her in his *heart.*"

By constant prayer do we attain the Kingdom, for Jesus said "With man it is impossible; but with God all things are possible."

Envy, hatred, ambition, covetousness, will destroy the capsule that contains the seed and thus corrupt the blood, as surely as sexual contact. Alcohol in all its deceptive forms is the arch foe to this life-seed and seeks by every means known to the enemy of man to destroy it. "No drunkard shall inherit the Kingdom of Heaven" because alcohol destroys the redeeming substance that enables man to understand or think in his heart the thoughts of the Spirit. Alcohol cuts the capsule that holds the Esse born every month in Bethlehem. Alcohol eats the fruit of the tree of life.

Gluttony is another enemy to regeneration. All excess of food, all that is not burnt up in the furnace—the stomach and intestinal tract, all that is not properly digested, ferments and produces acid which develops alcohol.

Auto-intoxication is common among those who overeat. Most everyone overeats.

The furnace, stomach and digestive tract becomes a distillery when the surplus food ferments, and thus becomes Babylon, the home of unclean birds and beasts which pander to carnal mind. Here we have the reason why sickness was considered Sin by the ancients. "To heal the sick and cast out devils" is the mission of the seed. "He that is born of God cannot sin, or be sick, for his seed remaineth in him." "The blood of Christ cleanseth from all sin," therefore from all disease. Here is the physiological explanation: When the *Christed* substance, the ointment from the river of Jordan, the oil in the spinal cord, reaches the pineal gland, it vibrates to a rate that causes new blood—the *new wine.* This is the blood of Christ that heals all infirmities. Unless so-called Christians repent of their sins, the doom of the church is at hand, *"Mene, mene teckel upharsin"* is written on the wall.

Here are the words that define a Christian: "These signs shall follow those who believe; they shall lay hands on the sick and they shall recover. They shall cast out devils and raise the dead. All the things that I do, ye shall do and greater things shall ye do."

If there be one Christian on earth today, let him stand forth and prove himself worthy. "He that overcometh, I will give to eat of the fruit of the tree of life." To overcome a habit is to *cease to do it.* When the earthly man is controlled by the spiritual man—the Lord God—he ceases to eat of the fruit, that is, *waste it.* This fruit is then carried up to the brain and "Eaten in the Father's Kingdom." "And the last enemy to be overcome is death." We *overcome* death by *ceasing* to die, and in no other way. "He that believeth in me, shall not perish." Those who die are sinners, and therefore are not Christians, for Christ Jesus was (is) without sin. "The wages of Sin *is death.*" Repent, forsake evil, take up thy Cross, call upon the Lord and He will abundantly pardon. "And the ransomed of the Lord shall return and come to Zion." When the sexual functions are used for the propagation of human bodies, there is no condemnation or sin. Motherhood is holy, pure, divine. But motherhood forced is crime. Unwilling motherhood has created the spirit of war and murder and well-nigh destroyed the race. Sexual union for pleasure alone is the broad road that leads to death. "And there shall be no more *Curse*"—Revelation. The word "Curse" has no reference to an oath. Curse means friction, to grind. The statement "Then Peter began to curse and swear" . . . And immediately the cock crew," when understood physiologically, fully explains the meaning of curse. Sexual commerce for the birth of children where the parents sacrifice themselves for their offsprings' sakes, or total abstinence, is written with a pen of flame on all the pages of ancient Scriptures and modern biology.

"And I saw a woman clothed with the Sun, having the Moon under her feet and twelve stars upon her head." The Sun is the "Son of Man," the product of her own body, saved and lifted up. The Moon refers to the generative life. Twelve stars are the twelve functions, typified by twelve zodiacal signs, which she has mastered through physical regeneration.

Miss Ruth Le Prade, the woman poet of the New Time, sings of the Kingdom as follows:

> "I am a woman free. My song
> Flows from my soul with pure and joyful strength.

It shall be heard through all the noise of things—
A song of joy where songs of joy were not.
My sister singers, singing in the past,
Sang songs of melody but not of joy—
For woman's name was Sorrow, and the slave
Is never joyful, tho he smiles.

I am a woman free. Too long
I was held captive in the dust. Too long
My soul was surfeited with toil or ease
And rotted as the plaything of a slave.
I am a woman free at last,
After the crumbling centuries of time;
Free to achieve and understand;
Free to become and live.

I am a woman free. With face
Turned toward the sun, I am advancing
Toward love that is not lust,
Toward work that is not pain,
Toward home which is the world,
Toward motherhood which is not forced,
And toward the man who also must be free.

With face turned toward the sun,
Strong and radiant-limbed,
I advance, singing,
And my song is as free
As the soul from which it flows.
I advance toward that which is, but was not;
I, the free woman, advance singing,
And with face turned toward the sun.
Let Ignorance and Tyranny
Tremble at the sound of my feet.

"When thou prayest, enter into thy closet and pray to thy Father in secret, and he shall reward thee openly."

The word Secret is derived from Secretions. The upper brain, the Cerebrum, contains the secretions, gray matter, creative or that creates, builds and supplies *all* the life force of the human temple,—Soul of Man's (Solomon's temple). Hence God, the Creator, dwells in you. The cerebellum is his throne. Prayer or desires expressed by man in the cerebellum for righteousness is answered in the cerebrum. Thus by prayer to God within, and in no other way, can man overcome the adversary or the "carnal mind which is at enmity to God."

Let us now consider Virgin Mary.

Virgin, pure.

Mar-y or mare, water.

Virgin Mary, pure water. Pure Sea. Pure Ether or spirit. Fish come out of the water. The water or fluids of the body give birth to the seed, or fish.

The Virgin Mary is not mentioned in the Allegory after the ascension of the Christed Jesus—the redeemed fish. Each person, male or female, must "work out their *own salvation.*"

All so-called sex reform that tolerates union of sexes, may be answered by:

> "There is a way that
> Seemeth right to man, ...
> The end of which is death."
> "In my Kingdom there is no marrying nor
> giving in marriage,
> But they are as the
> Angels in Heaven."

No page of the wonders of the human body—the temple of the living God—is more divinely scientific than the parable that follows:

> "The foolish man built his house on the sand
> And the rain washed it away."
> "The wise man built his house on a rock
> And it stood the storms, for it was
> builded upon a rock."

The Bible is a compilation of astronomical, physiological and anatomical symbols, allegories and parables.

In the technical terms of modern chemistry and physiology the above text is explained as follows: Sand and cement form rock or stone. Sand alone, without some medium—cement—is unstable, simply "shifting sand."

The Pineal gland, the *dynamo* that runs the organism of man, is composed of sand plus a cement, an ointment, a smear, found, as has been explained, in large quantity in the spinal cord, also to some extent, in all parts of the body. When this cement is wasted, as the Prodigal Son wasted his substance in riotous living, there being a deficiency of this precious oil, the pineal gland becomes brittle, and does not vibrate at a rate that vitalizes the blood and tissue at the health and strength rate, and the house, beth or body, falls.

In the common slang of the hour, we say: "He lacks the sand," or "grit."

The mineral salts of blood were called sand or salt by the Hebrews. The cell-salts that are found in the pineal gland are chiefly potassium phosphate, the base of the gray matter of the brain, but all of the 12 inorganic salts are represented. In Revelations, the pineal gland is called "the white stone." In Biochemistry, the phosphate of potassium is given as the birth salt of Aries people.

Those who build their house upon a rock are they who conserve the substance that unites with the sand—cell-salts—and thus form the rock upon which a body may be built that will be free from sin and sickness.

The mission of Jesus, the Christ, was to *triumph over death* and the grave, over matter, and transmute his body and also materialize at will. He not only succeeded in doing this, but stated most emphatically that all the things that he did, we may do also.

Did he proclaim the truth?

Answer, thou of little faith!

"Rock of Ages, cleft for me,
Let me hide myself in thee."

NOAH, THE ARK AND THE ANIMALS

FEW theologians are there, of to-day, who insist on a literal interpretation of the biblical story of the flood, Noah and the ark.

There are known to be 1656 species of mammals; 6266 species of birds; 642 of reptiles; 20 of oxen; 27 special of goats; 48 species of antelopes; insects, fish, turtles and creeping things on land and sea innumerable.

There is not a bit of geological evidence that the earth was ever totally submerged. But, going to the root of the words Noah, ark, Ararat, etc., it is quite easy to read the riddle of the allegory.

Noah is Hebrew for rest. Ararat simply means a mount or elevation. In English we say hill, mound, peak, mountain, etc. So in both Greek and Hebrew we find Nebo, Pisgah, Ararat, pinnacle of the temple, Zion, Gibeon, used to typify brain and pineal gland.

Ark, or boat, is used to symbol the seed (fish or Moses) born in the solar-plexus to be carried up through the regenerative process to the pineal gland. Moses was found in an ark and the ark of the covenant was carried by the children of Israel (see Jacob's 13 children) through the wilderness and *across* Jordan, where the "waters stood up at the City of Adam."

Adam means earth or sand. At the source of the spinal cord there is a body called medulla oblongata. Medulla means marrow or thick oil or ointment. This oblong body (oblongata in Latin) is a bed of mineral salts of the body and marrow. This precious oil (Christ) is received there by secretions from the cerebrum, the upper brain—the "Most High."

This oil flows down the spinal cord to the Caudia Equina, and this is a symbol of the Jordan and Dead Sea of Palestine.

Jordan means the "Descender" or oil flowing down. Witness: *Dove* or dive—to descend. Dove, i. e., a diver—"The Spirit of God descended like a dove, and a voice said, 'This is my Son,' " etc.

This occurred *after* the baptism of Jesus, the seed, in Jordan, the oil or Christ and the seed was then Christ-ened. The children of Israel (seeds born in solar plexus) are carried up the *strait* ("Narrow way that leads to life eternal,") known in modern physiology as vagus, or pneumo-gastric nerve. So, then, the poetical writer of this sublime epic could truthfully say that the twelve tribes (twelve mineral salts of the blood) "Set up stones" where they crossed Jordan, for the great nerve or "strait" crosses the medulla at the head or source of the spinal cord.

The animals taken up to Ararat, the pineal gland or "Pinnacle of the temple," simply means the transmutation of animal desires and propensities by saving the ark (seed) and crucifying it at Golgotha where it *Crosses* Jordan in medulla, the "Place of the Skull." (See cerebellum in anatomy and physiology.)

"And I saw a woman clothed with the Sun (Son)."

Woman the 4th Dimension.

The solar system has entered the "Sign of the Son of Man," where it will remain for over 2000 years. In astrology this sign is symboled as "The Water Bearer," while in Bible Alchemy it is represented by Dan, the fifth son of Jacob, and means "judgment," or "he that judges."

Uranus, the revolutionary planet, known as the "Son of Heaven," is now in the Aquarian sign of the "Son of Man," where it will remain until March, 1919.

From these statements it is easy to realize that all that is taking place in the world to-day is but a "working out" or a summing up of all that has been taking place for centuries.

The world is awakening, the old order is passing, worn-out traditions that are no longer applicable to present conditions must be replaced by new.

Radical and fundamental changes stare us in the face on all sides. Science, philosophy, religion, bodies politic and social—*all* are being shaken from their very foundations—to be rebuilt anew.

There is no equilibrium, no balance, no harmony, no equality, anywhere.

Nowhere do we see a better illustration of this unbalanced condition of the world than in man's attitude toward woman. For some time, now, this viewpoint has been gradually changing and Aquarian vibrations, or, in other words, the vibratory influence of the planets, have made conditions possible for this change.

Woman is at last coming into her own.

Co-equal with man! Mighty strides toward the regeneration of the human race will now be made.

With equilibrium of forces now possible world harmony shall grow apace.

All these truths can be mathematically expressed.

Four (4) means realization—one and three (1 plus 3) equals four.

Woman or mother comes from the Hebrew word Mem-M (womb, man, water, Mary—same meaning in all).

"I saw a woman clothed with the *Son,* the moon (from month-menses) under her *feet,* viz: the monthly seed born every twenty-eight and one-half days, thirteen yearly) twelve stars on her head." She controlled the twelve functions of the body. The Son signifies "Sun" or "Son of Man," the seed or product of her own life, saved and lifted up. The Moon refers to the generative life. Twelve stars are the twelve functions, typified by the twelve zodiacal signs which she has mastered through physical regeneration.

The seed or "Son-Sun" is born in the Solar Plexus in both man and woman every twenty-eight and one-half days.

Solar plexus, is the center of the body—the balance, or median line, and in Bible terminology means Bethlehem or manger. ("Beth" meaning house, and "lehem" meaning bread.)

The seed or "Son," born in this manger, or Bethlehem, is taken up the pneumo-gastric nerve to the base of the brain, Golgotha, or "place of the skull."

Here is the "cross," where the seed is increased in power a thousand fold, or, in other words, raised to a higher rate of vibration. This cross is made by the two very delicate nerves or wires, the Adi and Pingali.

Having been upon the cross, or having crossed over, the seed is Christed, and in the man or woman seeking to regenerate or "save," the seed is *saved),* it then enters the Optic Thalmus, the eye of the chamber, which "giveth light to all that are in the house," that is, to the twelve functions that are in the body, represented by the twelve signs of the zodiac.

Woman regenerated—"clothed with the Sun"—is the Queen of Sheba, in Bible symbology, and is represented by the number seven (7).

Then woman is Queen of 7. Sheba is seven in Hebrew, and Solomon's temple (soul-of-man) is the physical body where the Queen of Sheba found so many wonders.

Queen of seven *what?*

Man is only three (3) dimension.

Dimension means *line.*

The human body as well as the universe are geometrical figures, a fact which the old philosophers well knew, for they said that sound and number governed the laws of creation.

Man is proved to be a three dimensional creature by physiology; and woman is the fourth dimension, by the same means of proof.

In the thirty-first chapter of Jeremiah, twenty-second verse, we read: "A woman shall encompass a man."

Mathematically, a woman can encompass a man.

Man cannot encompass a woman, for he is only a three-line creature, while she is four. Therefore, four is able to encompass, or contain within its radius, three.

Woman may be represented by the square (four lines).

Man may be represented by the triangle (three lines).

Three and four do not balance, and never have. There has not been universal harmony or balance between them, for man has never considered woman his equal until very recently.

But man is "coming off his high horse," and the scales will soon balance.

All down the ages man has considered himself the "lord of all creation." The "spare rib" which he so condescendingly parted with in the so-called "beginning" unbalanced him entirely. He considered himself superior to woman and has continued to do so to the "end of the world," or "whorl of activity"—the activity or manifestation of the solar system in the last or previous sign, that representing the water age.

During the water age man conquered the water—inventions pertaining to water were perfected, etc., etc.

To return to the mathematical equation of man and woman:

The three dimensions or lines of man that can be shown on a physiological chart are the creative centers of the brain, the solar plexus and the sex organs. Woman also possesses the creative centers of brain, solar plexus and sex organs; *but* she also possesses another, and in a way the most wonderful of all—the *breast* that nourishes infant man. This is the fourth dimension or line. These imaginary lines are at equal distances from each other.

Work this out for yourselves on any physiological chart and you will never forget it.

In the triangle drawn to represent man we find the eye, also. This is a well-known Masonic symbol.

Man receives this from his mother, soul, Ego or mate, the feminine principle in the cosmos—from woman's fourth dimension.

There she encompasses him—i. e., gives him of her intuition. "Male and female created He (we) them." "Let us make man."

Yes, woman is coming into her own all over the world. What could man have done in the present world crisis without the aid of woman? And magnificently glorious has been her sacrifice, her utter forgetfulness of self. And it is not men, nor arms, nor might that will win this war, but sacrifice—the highest and the holiest sacrifice of which we can conceive, the sacrifice of woman.

TRANSLATIONS OF SCRIPTURE

"He that saveth his life shall lose it," etc., etc.
—*Mark: 8th Ch., 35th v.*

THE above sentence does not ring true and is not logical.

A Greek professor recently went to Oxford, England, for the sole purpose of looking into the Greek text in regard to this seeming inconsistency. (Also Luke 16:9. See below.)

The discovery was made that the letter N, (from nun, meaning a fish,) was omitted, also the letter O, and that a correct translation reads: "He that saveth his life shall *loosen* it," etc.

The seed, in the fable, or Jesus, said: "I am the way, the truth and the *life*," etc. Therefore, he that saveth his life (Seed) shall loosen it so that it may enter the "Strait and narrow way," etc. This strait is the Vagus Nerve. As has already been written, "I am the bread of life." Again, "Cast thy bread upon the waters"—i. e., the strait. Cast thy bread upon the water exactly harmonizes with "Loosen it."

Luke 16:9: "And I say unto you, make unto yourselves friends of the mammon of unrighteousness; that when ye fail they may receive you into everlasting habitations."

Literally the statement would nullify all the teaching of Jesus, and it is simply amazing that the so-called Christian world has so largely ignored it. However, a few critics from the orthodox ranks, not being at all satisfied with the rendering, have tried, in various ways, to reconcile the paradox, and to that end several pamphlets may be found in the theological departments of our colleges and universities.

Here is the explanation by a Greek scholar:

"Make unto your self *other* friends than those who worship the mammon of unrighteousness," etc.

Accepting the New Testament error, without question accounts for the great anxiety shown by churches of all denominations to secure the financial support of the wealthy, whether they be vital Christians, in belief, or nominally so. Proof of which may be seen in the end of the world, or age, nominally dominated by so-called Christianity.

Many worshipers of the mammon of unrighteousness exhibited much more horror over the destruction of costly cathedrals by the Huns than they did at the rape of women and slaughter of children by the Germans in Belgium, or murders by the sinking of the Lusitania.

Nothing can survive this "Day of Judgment" except it be founded upon the Truth, which liveth and reigneth forevermore.

JOSHUA COMMANDS THE SUN AND MOON TO STAND STILL

IN Eales' and Taber's Physiological Chart the solar plexus, a round body of tissue ganglion, may be plainly seen. Attached to this SUN (center) is a body called semi-lunar ganglion (half moon), which is attached to the vertebra and spinal cord. A median line (across the center of body) will divide these organs, half above the line, half below.

The upper halves of the sun and moon vibrate for spiritual man and the lower half for *natural,* or animal man.

"There is a natural and a spiritual body."—Paul.

Now Joshua, the seed, on its way to the pineal gland is made to say, "Sun, stand thou still on Gibeon."

Gibeon means a mound or elevation. So the seed (Joshua, a fish,) commands the animal vibrations of *solar* (sun) plexus to stand still, i. e., cease to continue to dominate the spiritual forces, "while I slay my enemies"—that is, the animal blood that predominates in carnal thought.

"And thou moon in the valley of Ajalon."

Ajalon means a "valley in Bethlehem," says a Bible dictionary.

Bethlehem—the house of bread: the *seed* is the bread.

Whoever conquers sex desire commands the sun and moon to stand still.

Who can do this?

"With man it is impossible, but with God all things are possible."

Therefore, all can succeed by asking help from the "Most High."

A cloud of witnesses may be found to substantiate the statement made above that the sun and moon in the Joshua story refer to the solar plexus and semi-lunar ganglion.

Eph. in Hebrew is prefix to many words meaning the centre or middle. It is defined in Smith's Bible Dictionary under the name Eph-ah, as "First in order of the sons of Midian, i.e., strife or contention between Michael and Apollyon occurs in the center of the body where the animal continually fights the upper force that seeks to lift up and regenerate the animal or natural man.

Ephah also means weight (measure or balance, Libra, the scales.)

Again, E-phes-dammin, "boundary of animal blood."

"I fought with wild beasts at Ephesus."—Paul.

Ephesians are the children of Ephesus, the solar plexus, therefore the seed. Paul the still small voice, or intuition, redeeming (lifting up).

The seeds constitute Paul's Epistle to the Ephesians.

Once more: "Eph-raim is joined to his idols; let him alone."

This epigram defines the physical man, "Dead in trespass and sin"—one who cannot be awakened by reasoning with him.

THE ANTI-CHRIST

PART I.

PRIMITIVE Christians, the Essenes, fully realized and taught the great truth that Christ was a substance, an oil or ointment, contained especially in the Spinal Cord, consequently in all parts of the body, as every nerve in the body is directly or indirectly connected with the wonderful "River that flows out of Eden (the upper brain) to water the garden."

The early Christians knew that the Scripture writings, whether written in ancient Hebrew or the Greek, were allegories, parables or fables based on the human body "Fearfully and wonderfully made."

These adepts knew that the Secretions (gray matter—creative) that issues (secretes) from the cerebrum was the *source* and *cause* of the physical expression called *man;* and they knew that the "River of Jordan" was symboled in the spinal cord and that the "Dead Sea" was used to symbol the Sacral Plexus at the base of the spinal column where the Jordan (spinal cord) ends typifying the entrance of Jordan into the Dead Sea.

The thick, oily and salty substance composing the Sacral Plexus, "Cauda Equina," (tail of the horse) may be likened unto crude petroleum, (Petra, mineral, or salt, and oleum—Latin for oil) and the thinner substance, oil or ointment in the spinal cord, may be compared with coal oil; and when this oil is carried up and *crosses* the Adi and Pingali, (two *fluid nerves* that end in a *cross* in medulla oblongatta where it contacts the cerebellum, Golgotha—*the place of the skull*) —this fluid is *refined,* as coal oil is *refined,* to produce gasoline—a higher rate of motion that causes the ascension of the airship.

When the oil (ointment) is crucified—(to crucify means

to *increase* in power a thousand fold—*not to kill*) it remains two days and a half (the moon's period in a sign) in the tomb (cerebellum) and on the third day ascends to the Pineal Gland that connects the cerebellum with the Optic Thalmus, the Central Eye in the Throne of God that is the chamber overtopped by the hollow (hallowed) caused by the curve of the cerebrum (the "Most High" of the body) which is the "Temple of the Living God" the living, vital substance which is a precipitation of the "Breath of Life" breathed into man—therefore the "Holy (whole Ghost" or breath.

The Pineal Gland is the "Pinnacle of the Temple." The modus operandi by which the oil of the spinal cord reaches the Pineal Gland is described in Part II.

Part II.

"There is no name under Heaven whereby ye may be saved except Jesus Christed and then crucified" (correct rendering of the Greek text).

Every twenty-eight and one-half days, when the moon is in the sign of the zodiac that the sun was in at the birth of the native, there is a seed or Psycho-Physical germ born *in,* or *out of,* the Solar Plexus (the Manger) and this seed is taken up by the nerves or branches of the Pneumo gastric nerve, (see Physiological Chart) and becomes the "Fruit of the Tree of Life," or the "Tree of good and evil"—viz: *good* if *saved* and "Cast upon the waters" (circulation) to reach the Pineal Gland; and *evil* if *eaten* or consumed in sexual expression on the physical plane, or by alcoholic drinks, or gluttony that causes ferment-acid and even alcohol in intestinal tract—thus "No drunkard can inherit the Kingdom of Heaven," *for acids and alcohol cut, or chemically split, the oils that UNITE with the mineral salts in the body and thus produce the monthly seed.*

This seed, having the odor of fish, was called Jesus, from Ichthos (Greek for fish) and Nun (Hebrew for fish) thus "Joshua the son of Nun." "I am the bread of life," "I am the bread that came down from heaven," "Give us this day our daily bread."

The fruit of the Tree of Life, therefore, is the "Fish-bread" of which thou shalt not eat on the plane of animal or Adam (earth; dust of the earth) plane: but to "Him that overcometh will I give to eat of the fruit of the Tree of Life" because he saved it and it returned to him in the cerebellum, the home of the spiritual man, the Ego.

The cerebellum is heart shaped and called the heart in Greek—thus "As a man thinketh in his heart so is he."

The bodily organ that men in their ignorance call heart is termed divider or pump in Greek and Hebrew. Our blood divider is not the button that we touch when we think, but it is the upper lobe of cerebellum that vibrates thought. The lower lobe is the animal (mortal) lobe that governs the animal world—that section of the body *below* the Solar Plexus, called lower Egypt—natural body—kingdom of earth—Appolyon—the Devil (lived, spelled backward) Satan (Saturn governs the bowels) etc.

Fire and brimstone (the lake of fire) comes from the fact that sulphur (brimstone) is the prime factor in generating the rate of motion called heat, and *overeating* develops a surplus of sulphur.

The Seed, born every twenty-eight and one-half days, making 13 in 365 days, that is 13 *months,* remains two and one-half days in *Bethlehem* (house of bread) then is carried up Pneumo (or vagus) gastric nerve and *across* the medulla oblongatta and enters the cerebellum to remain the two and one-half days, thus—"When Jesus was about *twelve* He appeared in the Temple teaching the doctors."

The age of puberty is about twelve. Then the *first born* seed appears and the sensation caused by its vibration tempts the native on the lower plane to do the thing that slays it, which is fully explained in Genesis by the serpent-sex desire-tempting Adam and Eve (allegorical characters). From Krishna to Moses and Jesus serpents and Pharaohs and He-rods have striven to slay the *first* born.

From the age of twelve to thirty in the life of Jesus nothing is recorded, for twelve refers to puberty, and 30 or 3 means physical, mental and spiritual, viz: body, (flesh or soul) fluids and Spirit (the Ego).

Breath is translated "soul" over 500 times in Bible, therefore soul is precipitated air (spirit) which may be lost in physical desire and expression (waste or (?), viz: to fall short) or saved by Regeneration. Read Matt. 17-28; also 1st Epistle of John—3-9.

So, at the age of 30, Jesus, the seed, began to preach to body, soul and spirit, and as the seed was (or is) descending the spinal cord, the substance of which is symboled by a formula of characters I. O. H. N. (as we symbol water by H_2O) it was baptized *of* John (not *by* John). Synonyms: Soul, John, Christ, Or (gold). *Jor*dan (word, Lord, oil, ointment).

Baptize is from the Greek Bapto, the *effect* of two chemicals when they unite and produce force that neither possessed singly. Here the seed, immersed in the oil, John, was so increased in power that "The Spirit of God descended like a dove and a voice out of Heaven said, 'This is my beloved Son' " etc.

Jordan means the *descender*—Dove, (to dive, a diver—see dictionary). Thus Jesus, the seed, was the son of man—the carpenter or builder—*until* it was baptized in the precious ointment that was secreted from the Most High (brain) and descended the spinal cord and was thus given power to start on its journey to Jerusalem (God's City of Peace) and to be crucified at Place of Skull, then remain two and one-half days in the tomb, and on the third day ascend to the Father.

As this seed consumes its force every twenty-eight and one-half days and another (born first) comes out of the Solar Plexus (Beth*lehem*), we see why he was (is) a "Sacrifice for our sins"; also we see that, as this seed, taking on the Christ oil, is enabled to reach the pineal gland and cause it to vibrate at a rate that *heals all manner of diseases*—that the statement "The blood of *Christ* cleanseth from *all* sin" or deficiencies viz: falling short of substance is literally true.

We also see why the statement was made "As ye see him go so shall he come again" and as this monthly going and coming occurs within the body the statement "Lo, I am with you always" rings true.

Part III.

During the first 300 years of the Christian era all that has been above written was understood by the real Christians, and about the end of that time the persecution of these

Essenes by the priesthood became so marked that they met in secret and always made the sign of the fish.

About the year 325, Constantine, the Pagan Roman Emperor, a monster in human form, like Nero, and the beast of August 1914, called the degenerate teachers of Christianity together at Nicea.

Constantine murdered his mother and boiled his wife in oil because they still held to the original doctrines of the Essenes. Constantine was told by the Priests of his time that there was no forgiveness for crimes such as his, except through a long series of incarnations; but the Anti-Christ sought to concoct a plan by which he hoped to cheat the Cosmic law.

And so it came to pass, after months of wrangling and fighting over the writings of the primitive Christians who clothed the wonders of the human body in oriental imagery, that the council, sometimes by a bare majority vote, decided which of the manuscripts were the "Word of God" and which were not.

The very important point in the minds of those ignorant priests—whether or no an angel had wings—was decided in favor of wings by three majority. The minority contended that, as Jacob let down a ladder for angels to descend and ascend upon that it was prima facie evidence that angels do not need wings.

Just think, for a minute, upon the collosal ignorance of these priests who did not know that Jacob in Hebrew means "heel catcher" or circle, and that ladder referred to the influence of the signs of the zodiac upon the earth; and as one sign rising every two hours forms a circle every twenty-four hours (the four and twenty Elders of Revelation) the outer stars of the rising suns (sons) "catching on" to the last sons (suns) of the sign ascending.

But now we come to the anti-Christ:

The Council of Nicea, dominated by Constantine, *voted* that the symbols of the human body were persons; that Jesus was a certain historical man, a contention utterly and indubitably without foundation, in fact, and that all who *believed* (?) the story would be *saved* and *forgiven, here,* and *now.* The idea appealed to the monster Constantine as an easy way out of his troubled mind and so the schemes of salvation by the actual blood of a real man or god was engrafted on the world.

Constantine and his dupes saw that the only way to perpetuate the infamy was to keep the world in ignorance of the operation of the Cosmic Law, so they *changed* "Times and seasons."

The date that they made the sun enter Aries was March 21st. Why? March 21st *should* be the *first* day of Aries, the head. April 19th should be the first day of Taurus, the neck, and so on through the twelve signs; but these schemers knew that thus expressing the truth the people might come to realize what was meant by "The heavens declare the glory of God." Again: the moon, in its *moon*thly round of 28½ days, enters the outer stars (or suns) of a constellation two and one-half days before it enters the central suns of the constellations that are known as the Signs of the Zodiac or the "Circle of Beasts." But even unto this day the whole anti-Christ world (so-called "Christian") except the astrologers, go by almanacs that make the moon enter a sign of the zodiac two and one-half days before it does enter it and thus perpetuate the lie of the pagan Constantine, the anti-Christ.

Let me close with a deadly parallel:

ANTI - CHRIST	CHRIST
Christ was a man born of woman. He died, and He will come again.	"Lo! I am with you always." "He that believeth" (believe means *to do*) "SHALL *never* die."
We are Christians and expect to *die* and then be *saved*.	"The wages of *sin is* death." "*All* that I do ye can do." "Be ye therefore *perfect* even as your Father is perfect."
Christ is greater than man, therefore can save us.	"Know ye not that the Holy Ghost" (breath) "dwelleth *in* you?" "The Kingdom of Heaven is *within you.*"
Only Jesus was conceived by the Holy Ghost.	"The earth endureth forever." "Thy will be done *in* earth as *in* heaven."

We must die in order to get into the "Kingdom." The earth will be destroyed.

"These SIGNS *shall* follow those who believe in me: they shall lay hands on the sick *and* they shall recover."

I am a Christian.

"He that is *born of God* will not sin, for *his seed remaineth in him.*"

I am born of God because I believe, or think, that a crucified saint, or good man, will save me from sin.

"Every spirit that confesseth that Jesus Christ *is come in the flesh* is of God."

"He that confesseth not that Jesus Christ *is* come in the flesh," etc., "*this is* the Spirit of the Antichrist."—John.

NOTE—*Is come,* the present or now. Was, or has, is not used in this statement.

The Greek and Hebrew texts of our Scriptures plainly teach that Jesus and Christ, John and baptism, crucifixion and ascension, the triumph of the Ego over the "Enemy death" are in the substance and potentialities of the body; and that they can and will save the physical body if conserved and not consumed (or wasted) in sexual or animal desire.

All of whatever name or religious denomination who teach a contrary doctrine agree with Constantine, who appeared in the "Latter days" of the Pure Christian Practice.

Who is the anti-Christ? Look at a world in ruins; does a good tree bring forth evil fruit?

The so-called teachers of and believers in Christianity believe as Constantine and his priests, that Christ is "out in the desert" of the Judean hills—out on Calvary. Do they ever look for the meaning of Calvary in Greek? Calvary means *a skull,* and Golgotha—the *place* of the skull, exactly where the seed is crucified.

One-half of the combatants in the world's Armageddon have been praying, as Constantine prayed, "for God's help for Christ's sake." The other half pray to the same imaginary God and Christ out in "The desert" of *their own ignorance* for "peace and victory."

Return and come into the God, and Christ *within you,* and the bugles will all sing truce along the iron front of war and the "Ransomed of the Lord will return to Zion with songs and everlasting joy upon their faces."

THE MYSTERY OF M. A. S. O. N. REVEALED

ALL symbols, allegories, parables or fables are founded in some basic principle.

The same may be said of great Epics, the book of Job, the twenty-two chapters of Revelation, or Milton's Paradise Lost.

Freemasonry is from "frea maso" to prepare, to knead, to materialize, as mixing and kneading flour, water, yeast, etc., forms bread or dough.

Mortar, made by mixing, etc., is, therefore, directly connected with the word mason in the world of art and labor.

Certain church organizations have opposed freemasonry and ostracized the members of that society very bitterly in the past. The occult reason will be briefly stated in the following:

There are twenty-two letters in the Hebrew alphabet. The twenty-first (falling short by one of completion) is Schin, from which the word sin is derived.

S, in the English alphabet, stands for Schin in Hebrew, as the characters of the ancient alphabet are reduced to one symbol in the modern.

Tay means a cross, +

Tau is used in many instances, even in the Bible (see 119th Psalm) but it is a palpable error, because there is no U in the Hebrew.

The Bible is a symbolical script of physiology and anatomy, a complete expose of the human body, written by men in dim past ages who knew a thousand times more about the "Fearfully and wonderfully made" temple of the Sol-of-man than modern physiologists who are just now beginning to perceive that the "Holy Ghost" (whole breath) really does reside in men and women, and the statement of the *oil, oint-*

ment, Christ, "Lo, I am with you always," is a physiological fact proven by the letters M. A. S. O. N.

M, or Mem, is the 13th letter of the Hebrew alphabet (also the 13th of the English alphabet) and means woman, Mary, water or mother.

A, the first letter, is Aleph, an ox, or male strength-father.

S, or Sin, here indicates that the woman and man or mother and father (Adam and Eve) *sinned* or *fell* short of something.

The early Christians, before the Anti-Christ, Constantine, appeared, understood the wonderful letters. Divine wisdom, seeing that M and A had fallen short (S) devised a plan to save the M and A.

O, from Ayin, the 16th letter, represents wisdom, or the All-seeing eye (see Optic Thalmus) the eye of the chamber in physiology. This is the eye of freemasonry, the third eye. "If thine eye be *single,* thy whole body will be full of light." "*Now* mine *Eye* seeth thee." Job. (33 degree member).

The All-seeing eye, O, here interposes and sends his son (S. O. N.) to save and supply the deficiency caused by the act that resulted in the *fall* when Eve and Adam ate the fruit.

NUN (really NON) from which we get N, is Hebrew for fish, the seed that is born (*first*) every 28½ days, or month, when the moon is in the sign of the zodiac that the sun was in at the birth of the native.

"Joshua, the son of Nun." "Joshua, Jehoshua, or Jesus" —see Smith's Bible Dictionary.

By saving this seed or *fish* (it has the odor of fish, from Ichthos, Greek for fish) it is "lifted up" carrying the precious ointment, and is thus *Christ*end, and crucified, refining it.

Thus Paul said, "If Christ be not raised or *crucified,* then is our preaching vain."

While I am credibly informed that but few Masons really understand their own symbols, the fact remains that they use them and that these perfect characters raise bitter antagonism in the churches.

Why?

May be because the sins of men and women can be "worked out"—"Work out your *own* salvation" or saliva, *salivation,* without the aid of priest or preacher.

This great truth is now, while the solar system is in Aquarius, "The sign of the Son of *Man* in the heavens," being so well recognized that the people will no longer suffer O, (All-seeing eye) and N (the Redeemer) to be cut off from M. A. S. and another Sin (S) added to Mas. for men and women to "celebrate."

M. A. S. O. N.

MASS.

Let the Lost Word be restored!

GLORIA IN EXCELSIS DEO!

"AS A MAN THINKETH IN HIS HEART SO IS HE"

THOUGHT is the creative power in the universe. Universal intelligence, operating as thought, sprang forth, "Spirit-sandalled and shod," at the appointed time and in the appointed place, and Lo! the planet earth, man's sorrowful star, became manifest.

Earth is man's sorrowful star for the reason that only by means of trouble and pain does humanity learn its lessons.

Spirit, manifesting on earth, uses earth as a negative pole, in order that the personality may grow. The mineral, vegetable and animal kingdoms use earth in much the same way. The earth is one plane of manifestation.

How can a man *think* in his *heart?*

The organ that divides blood was called by the ancients "dividing pump"—not heart. The real heart is the cerebellum and was so named by the Greeks and is the seat of thought.

Madam Blavatsky says, in the Secret Doctrine, that the cerebellum contains all, being the seat of intelligence.

The thinker, the individual or "man who never dies," has his home, therefore, in the cerebellum, under the shadow of the Almighty.

Read what the writer of the 91st Psalm has to say about this: "He that dwelleth in the secret place of the Most High shall abide under the shadow of the Almighty."

Secret (secretion, oil or ointment) place of the Most High—is that place where the secretion of oil or ointment is found. In the Bible we see so many references to oil-anointing, secret, secretions, etc.

This plainly shows that the place of the Most High is the cerebrum, that portion of the anatomy of man whence comes the oil or ointment—the precious substance that fructifies the brain of man and causes it to develop; it is that which nourishes the brain.

The abiding place of the Ego *is* "Under the shadow of the Almighty," since the cerebrum extends entirely over and around it.

And again the Psalmist says:

> "He will cover thee with his pinions
> and under his wings shalt thou take refuge."

The feathery convolutions which are plainly shown in the upper brain may be well compared to the feathers of a bird. The "Voice of the Silence" speaks of the Ego resting "Under the wings of the Great Bird."

The upper brain is composed of highly specialized substance. It is a reservoir of God's creative compounds. It is that God-making material—the Kingdom of Heaven wherein all is found.

"Seek ye first the Kingdom of Heaven and all things shall be added unto you."

"The Kingdom of Heaven is within you."

Heaven means "heaved up"—a high place.

The cerebrum is, then, the kingdom of heaven, for it is within us. By seeking it we draw from it the precious oil or ointment which shall nourish the brain and therefore cause it to grow and expand.

Certain parts of the brain cells are dormant. They are in a certain slow rate of motion or activity, and, therefore, answer to vibrations of their kind.

Let us suppose, for example, that little cell in the brain is composed of spirallae, spirals of nerves, seven sets of which can be seen by the trained occultist.

In a person of low development only three or four of these spirallae will be found to be active, while the man who is already working along the line of regeneration—living the life of self-sacrifice, will show five and six in active operation.

The higher and more lofty the *quality* of the thought, the finer or higher the vibration. Just as the vibration of the ether strikes upon the tympanum of the ear and produces sound—so are the spirallae of the brain cells operated upon by the fingers of the heavenly man, *when* the Kingdom is sought.

Thought, then, is a vibration, and as a man thinks so does he vibrate his brain cells.

How many people really think?

The war has done more to wake people up and set them to thinking than anything else ever could have done. It has started that process in many people—it has forced them to think.

Thought is a particular development of ideas, something entirely apart from the "hit-or-miss," "ramshackle" process which was supposed to be thought.

Let us begin to think; let us *choose* the material from which we shall build our temples—the temple of the "Living God."

The process that the average man calls thought is not consecutive thinking. God hasten the day when people will realize that all that is, has been or will be, is the result of thought.

Thought is both creative and destructive.

Not only are we making our bodies now, but we are making those which we shall wear in the future.

By the future I mean when the individual is reincarnated.

A great thinker has said: "Know this mighty fact, the soul is but the fruitage of thought tinctured and tarnished with the emotions, passions and desires of the flesh."

First, as regards the physical body. Thought selects the food by which the body is nourished. The cells of the body are being constantly destroyed and rebuilt. The purest food possible to obtain will construct a pure body. Vegetables, fruits and grains are of much finer construction than flesh, and hence can vibrate to much higher rates of motion.

Flesh is decaying animal matter and is detrimental to the highest development of man. Much meat eating thoroughly coarsens the body, and the marks of his calling are stamped on the face of the butcher.

Another example is that of a man who drinks. Alcohol brings about exactly the same results. The body cannot respond to any of the higher vibrations.

Just as surely as the note you strike on the piano must produce a certain tone, just so surely will your body answer to the same rate of vibration around it that it vibrates to in itself.

The high cost of meat during the present war has been a blessing in disguise, for it was the only means whereby people could be brought to realize that they could still *live* if they never ate meat. Then, after a time, they will begin to realize that they can enjoy much better health without it.

If you wished to do a fine, delicate piece of work, you would not use coarse or unwieldy instruments in doing it.

Just as true is it that the vehicle of the spirit—Solomon's Temple—must be delicately and finely constructed.

The body must be kept scrupulously clean and be given sufficient exercise.

If your body is not satisfactory to you, it is because you have indulged in thoughts that have marred its construction.

It is never too late to do *something* toward the reconstruction and regeneration of the body.

Start *NOW*.

The physical man is made up of twelve divisions, i.e. bone man, muscular man, nerve man, etc. These are all constructed with a certain cell salt or mineral as a base for each man or division of the body, see "Relation of the Mineral Salts of the Blood to the Signs of the Zodiac." Also "The Biochemic System of Medicine."

Each cell of the body is a *living,* throbbing intelligence. Each cell actually reaches out and grasps from out the water of life—that living stream of blood that is the life of the body—just the material it needs in its construction.

"The quality of the force called into action in any kingdom determines the quality of the offspring."

You are directly responsible for each thought that occupies your brain.

The soul is the thought man and the emotional man that occupies the physical body resembles it in form and feature. We do not here refer to the Spiritual Ego.

If, then, our thoughts build our bodies, what thoughts are the cell lives of the body filled with? We must naturally see that they are, in vast numbers, filled with thoughts of fear, strife and blood. Fears of microbes, disease, poverty, the neighbors, the weather, the night air, the dark, burglars, etc., etc.

Eternal strife for wealth, position and power, for material benefits. Benefits, so-called.

All this brings about war—the cell life gorged with blood, calling for the blood of its brothers.

Is not the cause of the war clear?

Do not thoughts pollute the very air? Is it not true that our thoughts affect those around us? What about the cells that we throw off from our bodies every minute—cells that we have built and that are impregnated with our thoughts?

What is the matter with the people in the world? For there is nothing the matter with the world itself.

Each cell, then, that we throw off from our bodies, hour by hour and day by day, bears the stamp of our thoughts upon it. These go to make up the record of our lives, which those whose eyes are opened can read. In occultism this is called the Akasic record.

Then each man is the recording angel.

"Like attracts like." Birds of a feather flock together." These are trite sayings.

We see, then, that the cellular construction and fineness of the tissues of the physical man is determined by the character of the thoughts we store away in them.

The prodigal son wasted his substance in riotous living. His thoughts were turned toward the indulgence of the lower passions, like the rich young man who went away sorrowful because he had many possessions. Therefore the precious substance, the oil or ointment, the elixir of life, was sold for a mess of pottage. The seed, or Christ, was not saved. If his thoughts had been pure and clean, the seed would have reached the cerebellum and would have increased in power a thousand fold. They then would have become the anointed of the Lord—would have received the oil or ointment. The prodigal would then have become the son "in whom the Father was well pleased."

When the thoughts of the disciple are purified from every undesirable thought—then he becomes the son of the Master for his thought flows like a river through the consciousness of his Lord.

His body has become transfigured, for each seed has become crucified and Christed. Each cell of his body has thrown off all its impurities and has become *white* in the blood of the lamb, for the blood of the lamb is as a crystal stream.

The process of regeneration causes the white corpuscles of the blood to overcome the preponderance of red, or Mars corpuscles.

Therefore the flesh becomes transparent—and he manifests more and more of the Father—he is no longer man—but has become a God.

Paul says: "Now, then, are we the sons of God."

"All things I have done ye can do, and greater."

As we go on living the renegerative life, the time comes when we no longer respond to any law within the physical realm, for all physical matter has been cast off from the body. "It is sown a material and is *raised* (because the seed has been raised—the rate of vibration has been raised) a *spiritual* body, and the Kingdom of Heaven has been attained.

> And now the sixth sense is opened,
> The seventh embraces the whole,
> And, clothed with the Oneness of Love,
> We reach the long-sought goal.
> And in all life's phases and changes,
> And along all the paths to be trod,
> We recognize only one power—
> One present, Omnipotent God.

HE THAT OVERCOMETH

The above sentence occurs nine times in Revelation.

To overcome a vice or habit means to cease to do it. In the Scriptures overcome is used to symbol the triumph of the Ego over sex or animal desire. It means the conquering of the carnal mind.

REVELATION, 2, 7—"He that hath an ear, let him hear what the Spirit saith unto the churches; To him that overcometh will I give to eat of the tree of life, which is in the midst of the paradise of God."

REVELATION 2, 11—"He that hath an ear, let him hear what the Spirit saith unto the churches; He that overcometh shall not be hurt of the second death."

REVELATION 2, 17—"He that hath an ear, let him hear what the Spirit saith unto the churches; To him that overcometh will I give to eat of the hidden manna, and I will give him a white stone, and in the stone a new name written, which no man knoweth saving he that receiveth it."

REVELATION 2, 26, 27—"And he that overcometh, and keepeth my works unto the end, to him will I give power over the nations." "And he shall rule them with a rod of iron; as the vessels of a potter shall they be broken to shivers; even as I received of my Father. And I will give him the morning star."

REVELATION 3, 5—"He that overcometh, the same shall be clothed in white raiment; and I will not blot out his name out of the book of life, but I will confess his name before my Father, and before his angels."

REVELATION 3, 12—"Him that overcometh will I make a pillar in the temple of my God, and he shall go no more out: and I will write upon him the name of my God, and the name of the city of my God, which is new Jerusalem, which cometh down out of heaven from my God: and I will write upon him my new name."

REVELATION 3, 21—"To him that overcometh will I grant to sit with me in my throne, even as I also overcame, and am set down with my Father in his throne."

REVELATION 21, 7—"He that overcometh shall inherit all things; and I will be his God, and he shall be my son."

A VISION OF IMMORTALITY AND THE NEW AGE

IT seemed to me that it was noon of a perfect day, and that I was wide awake. I stood upon a mountain top in Southern California and looked out to the West. I saw the clean page of the Balboa Sea and saw the white-maned horses of Neptune charge against San Pedro's seaward cliffs, only to be hurled back, but forever and aye returning, and charging again.

Far out I saw "Islands lift their fronded palms in air"—saw Santa Catalina holding up a hand from a sunken Empire, waiting for some geological palmist to read the story of an underseas nation.

The hills and crags of the delectable island were mirrored in the curling mists that rose and twisted about like things alive, and the mirage grew and spread until I fancied that the new Jerusalem was descending out of the heavens.

The Pacific Sea seemed like a smile of Infinite Love, and I heard the jubilant chorus of joyous Naiads.

In the foreground I saw the imperial City of Los Angeles, sitting on her hills of palms and olives, like Jerusalem of old, and I thought that hither might come the Queen of Sheba, as she came to King Solomon in the olden time, saying, "The half has not been told."

Eastward the Sierra Madre peaks lightly veiled their heads with mist and fleecy clouds as if to gently subdue their ineffable glory. I saw the clean trunked eucalipti, the pendant pepper boughs and the orange groves.

The desert blossomed and a commercial empire full-rounded spread out before me. From Santa Monica to San Diego broad boulevards marked the ocean beach and cliffs, castles and towers and temples were everywhere and the

"Voice of the Sea" chanted the Jubilee anthem of the victory of mind. The pathway of the eagle and the airship lay parallel, and man talked to man across the spaces without wire or artificial trasmitter or receiver; for the human brain had harnessed the Ether, and the telepathy of mind was regnant. Towns and cities had reached out friendly hands and altruism prevailed where selfishness had held sway. Architecture was uniform and stately. The animal instinct in man had evolved into human love and the miserable dwelling places called homes that offended aesthetic tastes in the days of poverty and competition had disappeared and in their places were Corinthian temples, frescoed, grained and gilded with gold.

I saw no locks on doors; no prisons.

In this land men and women did not hire out to another—did not sell their labor to the highest bidder, nor their souls for place, power or "distinction's worthless badge."

In this land of love and beauty there were no slaughter-houses—no stock pens.

The fruits and grains and vegetables, so bounteously yielded by Earth's breast, supplied the wants of the sons and daughters of the fair land of my vision.

I saw a race of people who worked and served for love, where co-operation had replaced competition, where love reigned instead of hate and envy.

In the land I saw, machines did the world's work, and all the people owned all the machines. They did not make machines out of flesh and blood and muscle, but from wood and iron and steel.

The people in this Arcadia knew that the "Heavens declare the glory of God" and that there is no language where their voice is not heard." These people knew that law is eternal and cannot be repealed or violated.

The products of labor in this happy land were distributed equally to all according to their needs, and there were no idle rich, no idle poor—no millionaires and no beggars.

Above the material world I saw miraged in the air Cabrillo's phantom ship, the mission at Old Town, San Luis Obispo, Soledad, Dolores, San Gabriel, Santa Barbara, San

Juan Capistrano, San Carlos, San Antonio, San Miguel, Los Angeles and San Juan. I saw the conquistadores, the Indian neophytes and heard their plaintive "Ave Marias." Some were counting beads and mumbling "Hail Marys" and some were weaving baskets. I saw the priests of the old time bow and kneel and make the sign of the cross; and in their midst, his face radiant as the sun, stood Junipero Serra, chanting "Gloria in Excelsis Deo." And then Ramona and Alessandro, hand in hand, looked down upon me and smiled as they walked along El Camino Real in the clouds.

And, from a yellow blossomed Acacia bough I heard a shuttle-throated mocking bird trilling liquid melody into the ears of Deity and I asked aloud "is not this the New Age? Am I immortal?"

And then a voice sweet as the voice of the Infinite Mother came out of the everywhere and I heard the words, "Yes, you are immortal. You stood in the rush of Divine Splendor when God said: 'Let there be Light.' You heard the morning stars chant the Epic of Creation. You saw the first procession of the constellations. You saw Orion light his clustering lamp out in the wilderness of the southern skies. You saw Arcturus rise from the unknown sea of silence and sentinel the Northern Pole. You saw the first rushing, blazing comet emerge from the awful realms of boundless space, sweep across measureless reaches of star dust, bearing upon its flaming front the glad intelligence that the rule of law is perfect: that suns, stars, systems follow the Cosmic urge and obey the Eternal Word.

You saw ancient Egypt, saw her magic, her rise, saw the pyramids built and the unnamable Sphinx sculptured to confound posterity. You saw Anthony and Cleopatra; and you saw Egypt hide her shame in the hot sands of the Nile.

You saw the ancient Briton throw off the yoke of feudalism; saw the French revolution and saw America banner the skies with the stars and stripes.

And if, in the operation of wisdom, the time shall come when the vast fabric of creation shall rock in universal spasm and totter to its fall, if the elements shall melt in fervent heat, the last sun die and the "Heavens be gathered together as a scroll"—yet thou, oh doubting one, shall stand erect, unafraid," and

"O'er the ruins smile,

And light thy torch again at nature's funeral pile."

And then I heard the bugles all sing truce along the iron front of war. I saw the battle flags furled. Soldiers returned to homes, shops, the fields, the orchards and gardens. Children laughed and women loved. The headsman and hangman retired and became forgotten horrors. Grass grew over battle trenches, flowers bloomed over deserted forts, vines clambered over arsenals and dreadnaughts rusted in the harbors. Earth was baptized with the golden light of Love.

With the "Eye behind the eyes" "I saw the Holy City beside the tideless sea" and I heard the angels striking all their harps of gold.

THE END